Molecular Biology Biochemistry and Biophysics
28

Alexander Levitzki

Quantitative Aspects
of Allosteric Mechanisms

With 13 Figures

Springer-Verlag Berlin Heidelberg New York 1978

Professor Dr. Alexander Levitzki
Department of Biological Chemistry
The Institute of Life Sciences
The Hebrew University
Jerusalem, Israel

ISBN-13:978-3-642-81233-0   e-ISBN-13:978-3-642-81231-6
DOI: 10.1007978-3-642-81231-6

Preface

The aim of this monograph is to summarize the essential features
which characterize the behavior of regulatory systems. Firstly
we discuss the laws which govern ligand binding in thermodynamic
terms. The basic cooperative and allosteric phenomena are described in thermodynamic terms without assuming any particular model. Then the molecular models developed by Monod, Wyman and
Changeux and by Koshland, Némethy and Filmer are presented in
detail. Special emphasis has been given to the analysis of the
Hill coefficient and its meaning both in thermodynamic terms and
in terms of the two allosteric models: the concerted model of
Monod, Wyman and Changeux and the sequential model of Koshland,
Némethy and Filmer. Special types of cooperativities are discussed in some detail namely, cooperativity stemming from ligand-
coupled protein association or dissociation, negative cooperativity and half-of-the-sites reactivity. A slightly extended space
was devoted to the discussion of negative cooperativity and half-
of-the-sites reactivity, since the existence of these phenomena
and their possible biological importance is less of a common
knowledge than positive cooperativity. This monograph does not
attempt to be a review of specific examples analyzed according
to one model or another. Rather, an attempt is made to provide
the reader with the quantitative tools to analyze any specific
regulatory system. Last but not least, I would like to thank
Prof. F.W. Dahlquist from the Institute of Molecular Biology,
the University of Oregon (Eugene) and Prof. D.E. Koshland, Jr.
from the University of California at Berkeley for providing me
with their unpublished work. I would also like to express my
deep gratitude to my collegue Mr. Y.I. Henis for carefully and
critically reviewing the manuscript and for his very construc-
tive suggestions.

Jerusalem, February 1978   A. LEVITZKI

Contents

Introduction

The study of the mechanisms which control cellular processes has become of prime importance in recent years. Three levels of regulatory control can be identified
a) genetic control, that is the control of gene expression;
b) hormonal control of cell function;
c) the control of regulatory enzymes.

These three types of control are involved in the entire spectrum of life processes encountered in the living cell. Genetic control refers to the switching on or switching off of genes; hormonal control refers to the switching on or off of metabolic processes, usually through a second messenger such as cAMP or Ca^{2+} ions; the regulation of enzyme activity usually involves cooperative enzyme-ligand interactions and the interaction of the enzyme with effector ligands which either switch on or switch off its activity. These regulatory mechanisms also apply to membrane-bound enzymes and receptor-dependent events which govern the communication between cell processes and the surrounding milieu.

In virtually all three categories of control listed above, the controlling event involves the interaction of a ligand with a specific receptor, usually a macromolecule. Following the binding step, a conformational transition is induced leading to the activation of a secondary signal. This statement can be illustrated by the following examples:

1. *Genetic Control*. The expression of the *lac* operon depends on the interaction of the β-galactoside inducer with the *lac* repressor bound to a specific site on the DNA. Upon binding of the β-galactoside to the repressor, the affinity of the latter to the DNA is decreased and the repressor is peeled off.

2. Hormonal Control. The interaction of adrenaline, glucagon, and a variety of other hormones with specific membrane receptors switches on the enzyme adenylate cyclase which catalyzes the production of cAMP, the "second messenger". In turn, cAMP initiates various biochemical processes, such as glycogenolysis, depending on the target cell. The interaction of norepinephrine with α-adrenergic receptors on the cell surface induces the insertion of Ca^{2+} through a specific "gate". In this case Ca^{2+} functions as the "second messenger" activating a variety of cellular processes such as K^{+} efflux from the cell.

3. Allosteric Control in Soluble Enzymes. The interaction of CTP with aspartate transcarbamylase (ATCase) inhibits the activity of ATCase. Analogously, the interaction of the effector GTP with CTP synthetase activates CTP synthetase.

These three types of control mechanism have one basis property in common: the interaction of a ligand with a specific receptor which triggers a certain biochemical event. It is generally accepted that conformational changes induced by these ligand-receptor interactions are transmitted specifically to the macromolecule or the subunit which is responsible for the biochemical function. This conformational change either switches on or switches off the biochemical signal, executed by the target macromolecule. This is generally known as subunit interaction and is the basis for the wide spectrum of regulatory mechanisms found in vivo. It is the molecular nature of these interactions which will be discussed in some detail in this monograph. The regulatory processes involving subunit interactions are best understood in soluble regulatory enzymes, and are less well understood in systems which involve membrane-bound receptors or in systems which involve protein-nucleic acid interactions. We shall therefore limit the detailed mathematical discussion to the analysis of subunit interactions of well-defined multi-subunit structures. The detailed understanding of subunit interactions in regulatory proteins is in fact the key to the understanding of more complex regulatory phenomena.

Chapter 1
Basic Concepts of Allosteric Control

In the late 1950s, a number of workers discovered that in bacteria, metabolic pathways which lead to the synthesis of essential metabolites are subject to feedback (or end-product) inhibition (Novick and Szilard, 1954; Umbarger, 1956; Yates and Pardee, 1956). It was established that, in many metabolic pathways, the terminal metabolite in the pathway functions as a specific inhibitor of the first enzyme in the pathway. Enzymological studies on a number of metabolic pathways revealed that the end-product, which is chemically distinct from the substrates of the initial enzyme in the pathway, inhibits the activity of the enzyme by binding to a site distinct from its active site. Since this feedback inhibitor is not *isosteric* with the substrate, the term *allosteric* effector was coined (Monod et al., 1963). It was established that the allosteric effector interacts with a specific *allosteric site* on the enzyme which is topographically distinct from the active site. The binding of the *allosteric effector* to the allosteric site brings about the *allosteric transition*, which consists of a specific conformational change at the active site (and other areas of the protein molecule), thus modulating its activity. It was very quickly realized that allosteric effectors are not necessarily inhibitors. They may also function as activators. In their classical paper Monod et al. (1963) give the example of phosphorylase b activation by 5'-AMP. In the latter case, 5'-AMP functions as a positive effector, switching on the phosphorylase reaction. It was therefore clear that allosteric effectors may be either negative effectors or positive effectors, depending on whether they inhibit or activate the reaction in question.

The essence of allosteric effects involves the interaction between the ligand-binding sites. When the site-site interactions

occur between chemically identical binding sites, one speaks of
an interaction between *homologous* sites. When the site-site interactions occur between chemically nonidentical sites, one speaks
of an interaction between *heterologous* sites. The binding of oxygen
to hemoglobin represents a case of homologous interactions,
whereas the inhibition of aspartate transcarbamylase by CTP represents a case of heterologous interactions. Homologous interactions are responsible for the phenomenon of *cooperativity*. If the
binding of a ligand molecule results in a change in the affinity
of the remaining sites towards the same ligand, the binding curve
obtained is nonhyperbolic, namely the process of ligand binding
cannot be described by a Langmuir or a Michaelis type of equation.
In the case of hemoglobin, the oxygen-binding curve is sigmoidal, since the binding of oxygen results in an increase of the
affinity of the remaining oxygen-binding sites towards oxygen.
This progressive increase in affinity is known as *cooperativity*
or, more accurately, *positive cooperativity*. As will be shown later,
the affinity of ligand-binding sites can in principle decrease
as a function of ligand saturation. This situation results in
negatively cooperative ligand binding. Today it is well established
that many regulatory enzymes also exhibit negative cooperatively
in ligand binding. Many of the regulatory enzymes which exhibit
cooperatively in ligand binding also possess allosteric sites
and, therefore, exhibit interaction between heterologous sites.
Monod et al. (1965) already pointed out that many allosteric enzymes such as ATCase (aspartate transcarbamylase) (Gerhart and
Pardee, 1963) and threonine deaminase (Changeux, 1961) bind the
substrate cooperatively. Furthermore, Monod et al. (1963) noted
that treatment of many allosteric enzymes with Hg^{2+} not only
desensitizes those enzymes towards their respective allosteric effectors, but also eliminates their substrate cooperative effect.
In other words, the desensitized enzyme binds the substrate molecule in a noncooperative (Michaelian) fashion. This observation indicated that cooperative interactions between identical
ligand sites (homologous sites) and the interactions between allosteric sites and active sites (heterologous sites) are linked
functionally. When the interacting sites are identical, the interactions are termed *homotropic* (example: binding of oxygen to

hemoglobin); when the interactions are among different types of sites, i.e., between active sites and regulatory sites, the interactions are termed *heterotropic* (e.g., the inhibition of ATCase by CTP).

Although originally allosteric proteins were defined as those proteins possessing regulatory sites distinct from their active sites, it is now a common practice to call a protein which exhibits site-site interactions an allosteric protein, even when the interacting sites are identical.

Chapter 2

The Structure of Multisubunit Proteins

I. General Principles

Regulatory enzymes are multisubunit structures. Structural analysis of regulatory enzymes has in every single instance revealed that the subunits interact with each other by way of specific noncovalent bonds. The subunits always form a well-defined geometrical structure, and the architecture obtained determines to a large extent the regulatory properties of the protein. It will therefore be necessary for us to discuss in some detail the principles of design of oligomeric proteins.

The terminology used to describe the structure of multisubunit enzymes is as follows: a protein which is composed of a number of subunits is called an *oligomeric protein*, or an *oligomer*. The subunits building the oligomeric structure are referred to as *protomers*, *monomers*, or *subunits*. The subunits (protomers) are bound to each other at specific intersubunit binding domains by noncovalent bonds. A subunit is usually composed of one polypeptide chain, although there are a limited number of cases where each subunit is composed of more than one polypeptide chain. Many regulatory proteins are composed of identical subunits such as glyceraldehyde-3-phosphate dehydrogenase, or from nonidentical subunits as in the cases of hemoglobin, tryptophan synthetase and aspartate transcarbamylase. In hemoglobin the α and the β subunits have identical functions. In other oligomeric proteins, such as aspartate transcarbamylase and tryptophane synthetase, the two types of subunit are different and also have different functions.

Most oligomeric proteins are composed of a small number of subunits which form closed oligomeric structures that rarely contain more than 12 subunits. In certain cases, such as multienzyme complexes or spherical viruses, the number of protomers is much larger. In their classical paper Monod et al. (1965) observed that the specificity of subunit-subunit recognition is so great that monomers of an oligomeric protein will associate exclusively with their identical partners even at high dilution and in the presence of other proteins. This principle has been verified by detailed renaturation studies on numerous oligomeric enzymes (Cook and Koshland, 1969). The existence of strong and specific noncovalent inter-subunit interactions which form these geometrically defined aggregates indicates that the subunit interact at specific binding domains. According to Monod et al. (1965), two modes of subunit interactions (Fig. 1) are possible:

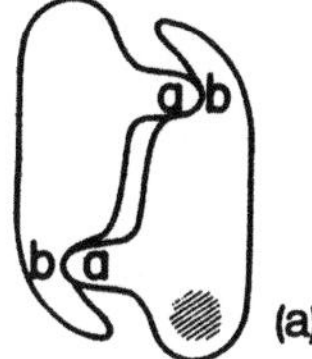

Fig. 1a and b. Modes of subunit association. (a) Isologous association, (b) heterologous association

a) Isologous association: the binding domain is made of two identical binding sets (Fig. 1a), each consisting of an "ab" contact. These binding domains are related to each other by a twofold rotational axis of symmetry.
b) Heterologous association: the binding domain is made of two binding sets (Fig. 1b) which differ from each other: one is a "bc" contact and the other is an "ad" contact. The arrangement of subunits in this case is also known as the head to tail mode of association.

The majority of proteins composed of subunits are either dimers or tetramers. The isologous mode of subunit interactions is the prevalent mode of aggregation in known protein dimers and tetramers. However, some heterologous tetramers such as tryptophanase (Morino and Snell, 1967) and pyruvate carboxylase (Valentine et al., 1966) do seem to have cyclic symmetry (C_4). Trimers, pentamers and hexamers, in which the mode of subunit aggregation is heterologous and therefore possess cyclic symmetry, are also known (Klotz et al., 1970).

II. Other Types of Protein Assemblies

The construction of oligomeric structures with more than four
subunits requires the use of both heterologous and isologous
subunit interactions. Thus, for example, an oligomer of point
symmetry 2:3, such as the dodecameric glutamine synthetase from
E. coli or *Salmonella* (Valentine et al., 1968) has three twofold
axes and four threefold axes of rotational symmetry. In this
case, the oligomer possesses one isologous binding set and two
heterologous binding sets. Point group symmetry 4:3:2 refers to
24 subunits arranged in cubic symmetry, and the highest point
group symmetry 5:3:2 relates to 60 subunits arranged in an ico-
sahedral symmetry, as is seen in the protein shells (capsids)
of spherical viruses. Higher protein assemblies which involve
closed geodesic, domelike arrangements are possible when one
allows the subunits to arrange themselves in such a way that
their environment is not always identical. This has been observed in large spherical viruses and has been termed quasi-
equivalence (Caspar and Klug, 1962). The complex assemblies are
beyond the scope of this monograph and will not be discussed
further. It should, however, be noted that some multienzyme
complexes such as pyruvate decarboxylase from bacteria and mam-
mals are arranged in the cubic point group symmetry (De Rosier
et al., 1971).

As shown in Figure 1, an isologous association of two subunits
has a twofold axis of rotational symmetry in the binding domain.
The isologous dimer can further associate to form an isologous
tetrahedral structure (Fig. 2), in which each subunit is attached

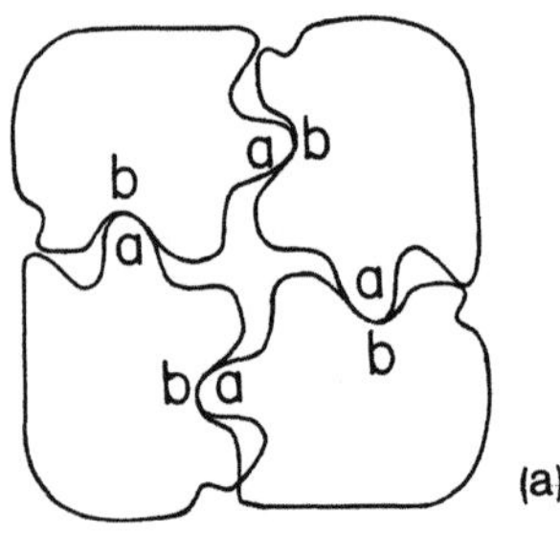

(a)

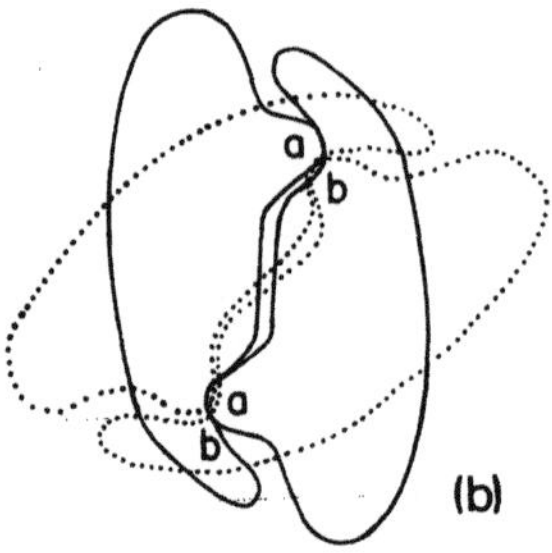

(b)

Fig. 2. (a) Heterolo-
gous and (b) isolo-
gous tetramer assem-
blies

to three other subunits via three types of isologous binding domains. The tetrahedral structure thus formed possesses three twofold axes of rotational symmetry. The situation in which tetrahedral symmetry exists can be schematically represented by Figure 3. Since the subunits are chyral themselves, no more elements of symmetry exist. This is why ping-pong balls are not good models for protein subunits. Indeed, all of the tetrameric enzymes composed of identical subunits which have been thus far analyzed by X-ray crystallography possess what is known as 2:2:2 point group symmetry, namely three twofold axes of rotational symmetry, as described above.

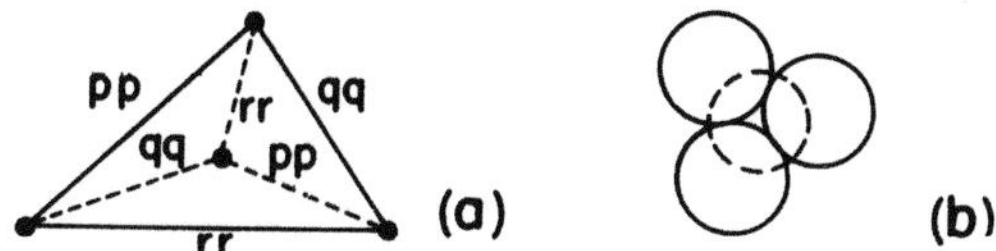

Fig. 3a and b. Schematic representation of tetrahedral symmetry. The four subunits are represented as spheres in (b). Each of the subunits makes three different types of contacts with its three neighbors. Three types of intersubunit domains can thus be identified: *pp*, *qq* and *rr*, each of which occurs twice. It can easily be seen that three twofold axes of symmetry occur in the structure

In an isologous dimer, the binding sets composing the binding domain are saturated within the dimer. The binding domains within a heterologous dimer are not internally saturated and will therefore lead to continued aggregation, forming polydisperse "open" structures (Fig. 1). Closed structures using heterologous binding domains can be formed once the subunits form rings. This is exemplified by the heterologous tetramer shown in Figure 2. In this heterologous tetramer no element of symmetry exists within the intersubunit binding domains. The symmetry in closed heterologous structures is a cyclic rotational symmetry, $\frac{360°}{n}$, where n is the number of subunits composing the closed heterologous structure. Thus, such subunit associations will lead to ring structures where the smallest closed heterologous structure is a trimer. Dimers would not be stable in a heterologous assembly (Fig. 1). It follows therefore that, when stable dimers are encountered as native species or as the dissociation products of a larger protein assembly, they must be isologous in nature. Deviations from the strictly isologous assembly are possible, and they will be discussed in Chapter 6.

In the following sections it will become apparent that a detailed knowledge of the quaternary structure of the protein is essential for the rigorous analysis of its interaction with ligands. The number of interacting surfaces between subunits is determined by their three-dimensional organization in space and any attempt to understand protein-ligand interactions involves the explicit knowledge of their geometrical arrangement. A more detailed discussion of the structure and symmetry of oligomeric enzymes can be found elsewhere (Klotz et al., 1970; Matthews and Bernhard, 1973).

Chapter 3

Cooperativity in Multisubunit Proteins – The Basic Concepts

I. The Hill Equation

The binding of oxygen to hemoglobin has intrigued scientists since 1904 when the sigmoidal nature of its oxygen-binding curves was first noted by Bohr et al. (1904). During the following seven decades of research, hemoglobin has played a key role in the different phases of understanding of allosteric phenomena.

In 1910 and 1913 Hill treated oxygen binding to hemoglobin as a single-step phenomenon (Barcroft and Hill, 1910; Hill, 1913). He suggested that hemoglobin is an aggregate of hemoglobin molecules $(Hb)_n$, where n is a specific but, at that time, an unknown integer. According to Hill, each Hb molecule possesses one oxygen-binding site, namely one iron moiety. The hemoglobin aggregate, according to Hill, binds oxygen according to the following equation:

$$Hb_n + nO_2 \rightleftharpoons Hb_n(O_2)_n \tag{1}$$

where the overall association constant for oxygen binding is given by:

$$K = \frac{[Hb_n(O_2)_n]}{[Hb_n][O_2]^n} \tag{2}$$

Hill suggested plotting $\log \dfrac{[Hb_n(O_2)_n]}{[Hb_n]}$ vs. $\log [O_2]$; this plot is known as the Hill plot. The ratio $\dfrac{[Hb_n(O_2)_n]}{[Hb_n]_{total}}$ is the fraction of hemoglobin oxygen-binding sites occupied by oxygen and is designated as $\bar{Y}$. It follows that the fraction of free hemoglobin is $1-\bar{Y}$. By rewriting Eq. (2) one obtains:

$$K = \frac{\bar{Y}}{(1 - \bar{Y})\,[O_2]^n} \tag{3}$$

and the Hill equation is therefore given by the expression:

$$\log \frac{\bar{Y}}{1 - \bar{Y}} = \log K + n\log[O_2] \tag{4}$$

According to Hill, a plot of $\log \dfrac{\bar{Y}}{1 - \bar{Y}}$ vs. the partial pressure of oxygen should yield a straight line with a slope n, a term known as the *Hill coefficient*. For oxygen binding to hemoglobin, Hill obtained the value of 2.8 for n.

According to the scheme suggested by Hill originally, the Hill coefficient should have been an integer. Hill explained his finding by stating: "In point of fact n does not turn out to be a whole number but this is due simply to the fact that aggregation is not into one particular type of molecule but rather into a whole series of different molecules, so Eq. (1) is a rough mathematical expression for the sum of several similar quantities with n equal to 1,2,3,4, and possibly higher integers".

Adair (1925) had established that hemoglobin is a molecule containing four equivalent binding sites for oxygen, and that the hemoglobin molecule does not dissociate in the absence of oxygen. Adair noted that the cooperative binding of oxygen was not accurately described by the Hill scheme, since Hill did not take into account the intermediate species $Hb(O_2)$, $Hb(O_2)_2$ and $Hb(O_2)_3$, but treated the oxygen binding as a single-step reaction [Eq. (1)].

Adair therefore wrote the binding steps for oxygen as a series of equilibria:

$$Hb + O_2 \overset{K_1}{\rightleftharpoons} Hb(O_2) \tag{5}$$

$$Hb(O_2) + O_2 \overset{K_2}{\rightleftharpoons} Hb(O_2)_2 \tag{6}$$

$$Hb(O_2)_2 + O_2 \overset{K_3}{\rightleftharpoons} Hb(O_2)_3 \tag{7}$$

$$Hb(O_2)_3 + O_2 \overset{K_4}{\rightleftharpoons} Hb(O_2)_4 \tag{8}$$

Each step is characterized by a thermodynamic equilibrium association constant K_1, K_2, K_3, K_4, respectively. These constants are in fact the Adair constants. The Adair equation is given by Eq. (9):

$$\bar{Y} = \frac{K_1[O_2] + 2K_1K_2[O_2]^2 + 3K_1K_2K_3[O_2]^3 + 4K_1K_2K_3K_4[O_2]^4}{4(1+K_1[O_2] + K_1K_2[O_2]^2 + K_1K_2K_3[O_2]^3 + K_1K_2K_3K_4[O_2]^4)} \tag{9}$$

The mono, di and tri bound hemoglobins are all included in this equation. The Hill equation, based on Hill's simplified treatment, predicts a straight line Hill plot over all the substrate concentration range with a Hill coefficient (slope) of 4.0. As is well known, the Hill plot is a straight line only in the neighborhood of $\bar{Y} = 1/2$ and the n value is 2.8 and not 4.0. In view of Adair's finding that hemoglobin does not undergo aggregation, Hill's original explanation for the coefficient being 2.8 had to be rejected.

It is now obvious that both the fact that the Hill plot is only linear within a limited ligand concentration range, and the fact that the Hill coefficient is frequently a fractional number are both due to the existence of intermediate species which should be taken into account in any analysis of a binding process. These are intermediate species in terms of the degree of saturation, not in terms of aggregation. Thus the Hill coefficient n in a real system is a function of ligand concentration and not a constant number as predicted by the simplified Hill approach.

The Hill equation [Eq. (4)] is still however widely employed[1] as a very useful diagnostic tool for estimating the degree of cooperativity, and in the determination of the minimal number of ligand-binding sites. It should be stressed that plotting ligand-binding data according to the Hill plot [Eq. (4)] is always a legitimate procedure. It is the interpretation of the Hill coefficient which requires reexamination and detailed analysis. The Adair approach is a general one and does not assume any molecular mechanism, since the equilibria considered [Eqs. (5-8)] are written on purely thermodynamic grounds. One should therefore emphasize that an Adair type of analysis of a binding process is always valid, and therefore extremely useful in the analysis of ligand binding to proteins. As will become apparent later, the molecular models developed to explain allosteric phenomena differ in the *interpretation* of the Adair binding constants.

From the discussion until now, it may appear that the Hill coefficient is a vague concept and should be used only as an auxilliary tool. This conclusion is, however, incorrect since it will be shown that the Hill coefficient can be expressed in terms of the binding constants of the Adair equation. Since the Hill coefficient is used extensively to describe the cooperativity of macromolecule-ligand interactions, we shall devote a whole chapter to the analysis of the Hill coefficient in terms of the Adair constants. It will be shown that the correlations between the Hill coefficient and the Adair parameters which will be derived enable one also to understand more fully the correct meaning of the Hill coefficient.

II. The General Adair Equation

Let us consider a protein possessing N identical ligand-binding sites. It is usually the case that each of the identical bind-

---

[1] "The equation originally deduced in 1910 from the aggregation theory had been laid decently to rest in the 1920s, its body lay mouldering in the grave, but apparently its soul goes marching on". A.V. Hill (1965).

ing sites resides on a separate subunit, where all the subunits are identical. All binding processes are fast and reversible, and no change in the molecular weight of the protein occurs during binding. Cases in which molecular weight changes are coupled to ligand binding will be considered separately (Ch. 6). K_1, K_2, K_i and K_N are the macroscopic thermodynamic association constants (the Adair constants). The binding of ligand to the oligomer can be described by the following equilibria:

$$E + X \rightleftharpoons EX \qquad \psi_1 = \frac{[EX]}{[E][X]} \qquad \text{where } \psi_1 = K_1 \qquad (10)$$

$$E + 2X \rightleftharpoons EX_2 \qquad \psi_2 = \frac{[EX_2]}{[E][X]^2} \qquad \text{where } \psi_2 = K_1K_2 \qquad (11)$$

$$E + 3X \rightleftharpoons EX_3 \qquad \psi_3 = \frac{[EX_3]}{[E][X]^3} \qquad \text{where } \psi_3 = K_1K_2K_3 \qquad (12)$$

$$E + iX \rightleftharpoons EX_i \qquad \psi_i = \frac{[EX_i]}{[E][X]^i} \qquad \text{where } \psi_i = K_1K_2K_3 \ldots K_i \qquad (13)$$

$$E + NX \rightleftharpoons EX_N \qquad \psi_N = \frac{[EX_N]}{[E][X]^N} \qquad \text{where } \psi_N = K_1K_2K_3 \ldots K_i$$
$$\ldots K_N \qquad (14)$$

the quantities ψ_1, ψ_2, ψ_3, ψ_i, and ψ_N are the formation constants for the corresponding complexes. The average number of ligand molecules bound per mol of protein is given by:

$$N_X = \frac{[EX] + 2[EX_2] + 3[EX_3] + \ldots + i[EX_i] + \ldots + N[EX_N]}{[E] + [EX] + [EX_2] + [EX_3] + \ldots + [EX_i] + \ldots + [EX_N]} \qquad (15)$$

where N_X is the average number of molecules X which bind to the protein.

$$N_X = \frac{\psi_1[X] + 2\psi_2[X]^2 + 3\psi_3[X]^3 + \ldots + i\psi_i[X]^i + \ldots + N\psi_N[X]^N}{1 + \psi_1[X] + \psi_2[X]^2 + \psi_3[X]^3 + \ldots + \psi_i[X]^i + \ldots + \psi_N[X]^N} \qquad (16)$$

which can be written as:

$$N_X = \frac{\sum_{i=1}^{N} i\psi_i [X]^i}{1 + \sum_{i=1}^{N} \psi_i [X]^i} \qquad (17)$$

Equation (17) is a generalized form of the Adair equation [Eq. (9)] written originally for the four binding steps of oxygen to hemoglobin.

III. The Statistical Correction

In order to obtain the correct free energy of association for each microscopic binding step to each of the binding sites, a statistical correction must be applied to account for the number of identical binding sites on the protein. These statistical corrections are related to the coefficients of the binomial expansion. If the ligand X binds to the species EX_{i-1} to form EX_i which possesses the total of N sites, there are N-i+1 ways for the ligand to get on the protein and i ways to dissociate. Thus the macroscopic (thermodynamic) association constant is related to the intrinsic (microscopic) association constant by the following expression:

$$K = \frac{N - i + 1}{i} K' \qquad (18)$$

where K is the macroscopic thermodynamic association constant (the Adair constant) and K', the intrinsic (microscopic) association constant. K' is the exact measure for the affinity of the ligand to the individual binding site.

IV. The Hill Coefficient in Terms of Intrinsic Ligand Affinities

We have already seen that the Hill coefficient, n, in real systems is not a constant number, and is in fact a function of ligand concentration. It is therefore pertinent that a quantitative interpretation of the Hill coefficient, n, obtained from the plot of $\log \frac{\bar{Y}_x}{1-\bar{Y}_x}$ versus $\log [X]_{free}$, where $\bar{Y}_x$ is the fraction of sites occupied and $[X]_{free}$ is the concentration of free ligand at which the value of $\bar{Y}_x$ is attained. Since the chemical potential of the free ligand is proportional to $\log[ligand]_{free}$, it is apparent that the Hill plot actually examines the dependence of the difference between the chemical potential of the bound state ($\log \bar{Y}_x$) and the free state, $\log(1-\bar{Y}_x)$, on the chemical potential of the ligand. A more detailed thermodynamic interpretation of the Hill coefficient has been given by Wyman (1964, 1968) and a statistical mechanical evaluation of the Hill coefficient has also been published (Heck, 1971). However, in our present survey we will be more interested in evaluating the meaning of the Hill coefficient in terms of the intrinsic affinities of the ligand to the protein. Once this is achieved, the interpretation of the Hill coefficient according to the parameters defined by the different allosteric models will be considered.

V. The Hill Coefficient at 50% Ligand Saturation

Experimentally it is usually found that binding data are easily obtained at a ligand concentration yielding 50% saturation. The concentration of ligand yielding 50% saturation is also a useful quantity in describing the average affinity of the binder under study. It has therefore become a matter of practice that the Hill coefficient is measured at 50% saturation. Mathematically, as will become apparent, one is able to obtain useful expressions for the Hill coefficient at 50% ligand saturation. The Hill coefficient, n, at free ligand concentration yielding 50% saturation ($\bar{Y}_x = 1/2$) is known as n_H. Let us derive the general formula for the Hill coefficient n_H at 50% ligand saturation.

From Eqs. (15) and (16) one can obtain the general expression for $\bar{Y}_x$:

$$\bar{Y}_x = \frac{Nx}{N} = \frac{1}{N} \, x \, \frac{K_1[X]+2K_1K_2[X]^2+3K_1K_2K_3[X]^3+\ldots+iK_1K_2K_3\ldots K_i[X]^i}{1+K_1[X]+K_1K_2[X]^2+K_1K_2K_3[X]^3+\ldots+K_1K_2K_3\ldots K_i[X]^i+\ldots}$$

$$\frac{+NK_1K_2K_3\ldots K_i\ldots K_N[X]^N}{+K_1K_2K_3\ldots K_i\ldots K_N[X]^N} \tag{19}$$

or

$$\bar{Y}_x = \frac{Nx}{N} = \frac{1}{N} \, x \, \frac{\psi_1[X]+2\psi_2[X]^2+3\psi_3[X]^3+\ldots+i\psi_i[X]^i+\ldots+N\psi_N[X]^N}{1+\psi_1[X]+\psi_2[X]^2+\psi_3[X]^3+\ldots+\psi_i[X]^i+\ldots+\psi_N[X]^N} \tag{20}$$

where the K values are the thermodynamic (statistically uncorrected) binding constants. The independent parameters in Eq. (20) are in fact the ψ values. Equation (20) can be written in the general form of:

$$\bar{Y}_x = \frac{1}{N} \, x \, \frac{\displaystyle\sum_{i=1}^{N} i\psi_i[X]^i}{1 + \displaystyle\sum_{i=1}^{N} \psi_i[X]^i} \tag{21}$$

The expression for the Hill coefficient is:

$$n = \frac{d\ln \dfrac{\bar{Y}_x}{1-\bar{Y}_x}}{d\ln[X]} = \frac{1}{\bar{Y}_x(1-\bar{Y}_x)} \, x \, \frac{d\bar{Y}_x}{d\ln[X]} \tag{22}$$

At the midpoint ($\bar{Y}_x = 1/2$) Eq. (22) obtains the form:

$$n = n_H = \frac{d\ln \dfrac{\bar{Y}_x}{1-\bar{Y}_x}}{d\ln[X]} = 4 \frac{d\bar{Y}_x}{d\ln[X]} = 4[X] \frac{d\bar{Y}_x}{d[X]} \tag{23}$$

It should be emphasized that Eq. (23) was arrived at without any assumption concerning the nature of ligand binding. Using Eqs. (21) and (23) a general expression for n_H can be obtained:

$$n_H = \frac{4}{N} \times \frac{\sum_{i=1}^{N} i^2 \psi_i [X]^i \left(1 + \sum_{i=1}^{N} \psi_i [X]^i\right) - \left(\sum_{i=1}^{N} i\psi_i [X]^i\right)^2}{\left(1 + \sum_{i=1}^{N} \psi_i [X]^i\right)^2}$$

$$= \frac{4}{N} \left(\frac{\sum_{i=1}^{N} i^2 \psi_i [X]^i}{1 + \sum_{i=1}^{N} \psi_i [X]^i} - \frac{\left(\sum_{i=1}^{N} i\psi_i [X]^i\right)^2}{\left(1 + \sum_{i=1}^{N} \psi_i [X_{0.5}]^i\right)^2} \right) \tag{24}$$

However at $\bar{Y} = 1/2$ one can show from Eq. (21) that:

$$\frac{N}{2} = \frac{\sum_{i=1}^{N} i\psi_i [X_{0.5}]^i}{1 + \sum_{i=1}^{N} \psi_i [X_{0.5}]^i} \tag{25}$$

where $[X_{0.5}]$ is the free ligand concentration at 50% saturation. Inserting Eq. (25) into Eq. (24) yields a general expression for the Hill coefficient at 50% saturation:

$$n_H = \frac{4}{N} \left(\frac{\sum_{i=1}^{N} i^2 \psi_i [X_{0.5}]^i}{1 + \sum_{i=1}^{N} \psi_i [X_{0.5}]^i} - \frac{N^2}{4} \right) \tag{26}$$

VI. The Maximal Hill Coefficient

When the cooperativity of the system is maximal, the enzyme is either in its completely liganded state EX_N or in its ligand-free state. This situation is in fact the one assumed by Hill originally (Sec. I). Under these conditions, Eq. (21) obtains the form:

$$\bar{Y}_x = \frac{1}{N} \times \frac{N\psi_N [X]^N}{1 + \psi_N [X]^N} \tag{27}$$

Since $\sum_{i=1}^{N} i\psi_i [X]^i = N\psi_N [X]^N$ and $\sum \psi_i [X]^i = \psi_N [X]^N$, it is also clear that under these conditions $\psi_N [X_{0.5}]^N = 1$. Inserting into Eq. (26) the relationships $\sum_{i=1}^{N} i^2 \psi_i [X_{0.5}]^i = N^2 \psi_N [X_{0.5}]^N$ and $\sum_{i=1}^{N} \psi_i [X_{0.5}]^i = \psi_N [X_{0.5}]^N = 1$, one obtains that $n_H = N$, namely, the Hill coefficient equals the total number of interacting sites.

This proof was originally provided by Wyman (1964) and by Weber and Anderson (1965). Indeed the Hill coefficient is always a measure for the minimal number of interacting sites, and its maximal value never exceeds this number.

VII. The Limiting Values of the Hill Slope

It was already pointed out by Wyman (1964) that at very low and very high ligand occupancy, the slope of the Hill plot tends to unity (Fig. 4). These regions of the Hill plot correspond to the

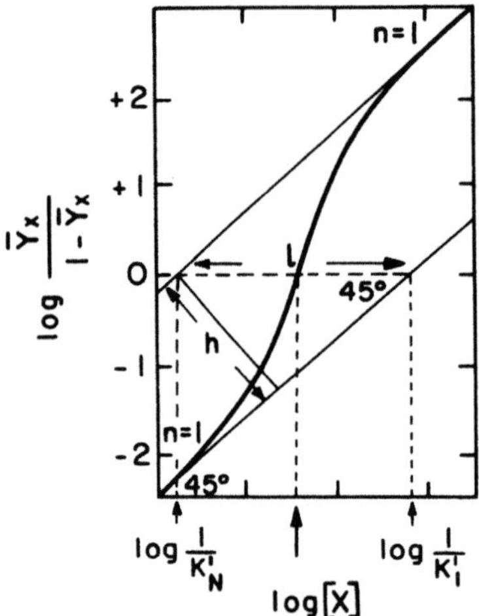

Fig. 4. The Hill plot. The interaction energy in an n-mer. The Hill plot is shown to be nonlinear. The limiting Hill slopes at vanishing ligand concentration ($[X] \to O$) and at infinite ligand concentrations ($[X] \to \infty$) is 1.0, as was proved by Wyman (1964). The slope at the midpoint ($[X] = [X_{0.5}]$, position marked with an *arrow*) is defined as n_H and is the quantity of interest. The *straight lines* obtained by taking the limiting Hill slopes at low and high ligand concentrations can be used to obtain directly the values of the ligand affinity to the first site ($\frac{1}{K'_1} = K'_1$) and the last site ($\frac{1}{K'_N} = K_N$)

binding of the first and last ligand respectively. Extrapolation of these limiting slopes allows the calculation of the apparent binding constants for the first and the last ligand respectively (see Fig. 4).

VIII. The Allosteric Dimer

The analysis of the behavior of a dimer is the simplest, since it is the smallest oligomeric structure which can possess co-operative interactions. It will also be shown that oligomeric structures composed of isologous dimers, and in which subunit interaction occur only within pairs obey the relatively simple equations governing dimer behavior. In general, ligand binding to a dimer can be described according to the following scheme:

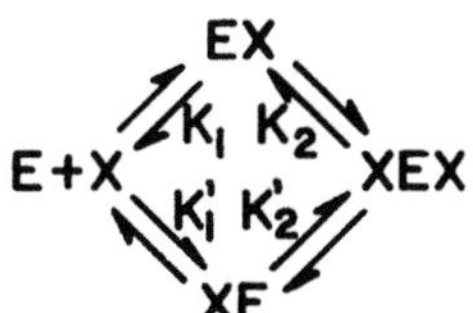

Fig. 5. The general scheme of ligand binding to a dimer. A protein dimer E is capable of binding the ligand X by two alternative pathways. K_1, K_2, K_1', K_2' describe the ligand affinities at the various steps

where the K values are the binding constants. K_1 and K_1' refer to the respective affinity of the two sites and toward the first binding ligand, where K_2 and K_2' are the affinities for the second binding ligand. From Figure 5, it is apparent that the relationship:

$$K_1 K_2 = K_1' K_2' \tag{28}$$

must hold.

Because of the fact that $\psi_1 = K_1 + K_1'$ and $\psi_2 = K_1 K_2$, the binding function which measures the number of ligands bound per protein for this case is given by:

$$N_X = \frac{\psi_1 [X] + 2\psi_2 [X]^2}{1 + \psi_1 [X] + \psi_2 [X]^2} = \frac{(K_1 + K_1')[X] + 2K_1 K_2 [X]^2}{1 + (K_1 + K_1')[X] + K_1 K_2 [X]^2} \tag{29}$$

Equation (29) applies to the most general dimer case and is easily obtained from the general Eq. (21). Two special cases can be distinguished: (a) a symmetric dimer where $K_1 = K_1'$ and therefore $K_2 = K_2'$. The affinity towards the ligand changes from K_1 to K_2 due to conformational changes brought about by the binding of the first ligand; (b) the second case is the case of the asymmetric dimer where $K_1 \neq K_1'$, but where $K_2' = K_1$ and $K_1' = K_2$ in the scheme shown above (Fig. 5). In this case, no interaction between the sites occurs.

The $[X_{0.5}]$ value for the dimer case is obtained from Eq. (29), solving for $N_X = 1$ ($\bar{Y}_X = 1/2$). One obtains:

$$[X_{0.5}] = \psi_2^{-\frac{1}{2}} = \frac{1}{(K_1 K_2)^{\frac{1}{2}}} = \frac{1}{(K_1' K_2')^{\frac{1}{2}}} \tag{30}$$

The Hill coefficient at $\bar{Y}_X = 1/2$ is then:

$$n_H = \frac{4}{\dfrac{\psi_1}{(\psi_2)^{\frac{1}{2}}} + 2} \tag{31}$$

This expression can be arrived at from the general equation [Eq. (26)] by inserting i = 2, and using Eq. (30).

Equation (31) is a general expression for the Hill coefficient in the dimer case. Two cases can be considered.

a) The symmetric dimer: In this case the two binding sites are identical in the absence of ligands. Therefore, $\psi_1 = 2K_1'$ and $\psi_2 = K_1' K_2'$ and

$$n_H = \frac{2}{1 + \sqrt{\dfrac{K_1'}{K_2'}}} \tag{32}$$

where K_1' and K_2' represent the intrinsic ligand affinities to the first and the second site respectively.

It is immediately seen that both negative cooperativity ($n_H < 1$), noncooperativity ($n_H = 1$), and positive cooperativity ($n_H > 1$) are predicted.[2]

b) The asymmetric dimer with $K_2' = K_1$ and $K_1' = K_2$ (Scheme 5): In this case Eq. (31) assumes the form:

$$n_H = \frac{4}{2 + \sqrt{\dfrac{K_1'}{K_2}} + \sqrt{\dfrac{K_1''}{K_2'}}} \qquad (33)$$

In this case n_H can assume values only equal or below 1.0 ($n_H < 1$) since $(\sqrt{\frac{a}{b}} + \sqrt{\frac{b}{a}}) \geqslant 2$. Thus the pre-existing noninteracting dimer model is only able to account for noncooperativity and negative cooperativity, and not for positive cooperativity. This is true since in the asymmetric dimer the relations $K_1 > K_2$ and $K_1' < K_2'$ must hold.

In experimental cases where negative cooperativity occurs in binding, both (a) and (b) can fit the data exactly. The reason for this is because the binding equation for both models possesses only two independent parameters, namely ψ_1 and ψ_2.

IX. The Multi-Dimer Case

If m is the number of dimers in the protein molecule, the binding equation is given by:

$$N_X = \frac{m(\psi_1[X] + 2\psi_2[X]^2)(1 + \psi_i[X] + \psi_1[X]^2)^{m-1}}{(1 + \psi_1[X] + \psi_2[X]^2)^m} \qquad (34)$$

---

[2] Negative cooperativity refers to a situation where the ligand affinity progressively decreases as a function of ligand occupancy, and positive cooperativity refers to a situation where the ligand affinity increases as a function of ligand occupancy (see Ch. 1).

where m is the number of dimers in the protein assembly. Equation (34) reduces to:

$$N_X = \frac{m(\psi_1[X] + 2\psi_2[X]^2)}{1 + \psi_1[X] + \psi_2[X]^2} \tag{35}$$

Equation (35) is identical to Eq. (29) obtained for the dimer case, multiplied by m, the number of dimers in the oligomer. Both the dimer equation [Eq. (29)] and the multiple dimer case [Eq. (35)] yield the same saturation function $\bar{Y}_X$:

$$\bar{Y}_X = \frac{1}{2} \times \frac{\psi_1[X] + 2\psi_2[X]^2}{1 + \psi_1[X] + \psi_2[X]^2} \tag{36}$$

Therefore the expressions for the 50% ligand concentration (midpoint concentration) $[X_{0.5}]$, and the Hill coefficient at 50% saturation for the multidimer case are identical to those derived for the simple dimer case.

X. The Allosteric Tetramer

Only in the case of a dimer can one obtain exact analytical equations for the Hill coefficient. One cannot derive an analytical expression for the case of a protein tetramer by inserting i = 4 in Eq. (26). However, under conditions where the ligand-binding curve plotted as $\bar{Y}_X$, versus log[X] or the Hill plot is symmetric about the midpoint ([X] = $[X_{0.5}]$), one can solve Eq. (26) and obtain an expression for the Hill coefficient at 50% saturation. The symmetry condition states that:

$$\bar{Y}_{A[X_{0.5}]} + \bar{Y}_{\frac{1}{A}[X_{0.5}]} = 1.0 \tag{37}$$

where A is any positive number. The condition of symmetry results in a specific relationship between the Adair constants. This relationship is:

$$\psi_1^2 \, \psi_4 = \psi_3^2 \tag{38}$$

If the binding sites are chemically identical, Eq. (38) is transformed into:

$$K_1 K_4 = K_2 K_3 \quad \text{or} \quad K_1' K_4' = K_2' K_3' \tag{39}$$

where K_1, K_2, K_3 and K_4 are the intrinsic dissociation constants and K_1' through K_4' are the intrinsic association constants.

The half-saturation point is defined by the following relationship:

$$[X_{0.5}] = (\psi_4)^{-\frac{1}{4}} = (K_2' K_3')^{-\frac{1}{2}} = (K_1' K_4')^{-\frac{1}{2}} \tag{40}$$

A proof for Eq. (40) is given in the Appendix.

Let us define:

$$a = \frac{K_2}{K_1} = \frac{K_1'}{K_2'} \tag{41}$$

and

$$b = \frac{K_3}{K_2} = \frac{K_2'}{K_3'} \tag{42}$$

It follows that

$$[X_{0.5}] = (a^2 b)^{\frac{1}{2}} (K_1')^{-1} \tag{43}$$

From Eq. (40) and Eq. (26) one can obtain an expression for the Hill coefficient at 50% saturation:

$$n_H = \frac{4[(a^2 b) + 1]}{4(a^2 b)^{\frac{1}{2}} + 3ab + 1} = \frac{4\left(\dfrac{K_1'}{\sqrt{K_2' K_3'}} + 1\right)}{\dfrac{4K_1'}{\sqrt{K_2' K_3'}} + 3\dfrac{K_1'}{K_3'} + 1} \tag{44}$$

where K_1', K_2', K_3' and K_4' are intrinsic binding constants, or

$$n_H = \frac{4\left(\sqrt{\dfrac{K_1'}{K_4'}} + 1\right)}{4\sqrt{\dfrac{K_1'}{K_4'}} + 3\dfrac{K_2'}{K_4'} + 1} \tag{45}$$

Two extreme cases can be considered using Eq. (44):

1. The case of noninteracting sites where a = b = 1. In this case $n_H = 1.0$, as expected.

2. The case of infinite cooperativity where a = b = 0. In this case $n_H = 4.0$, as is also expected.

XI. The General Tetrameric Case

It is clear that in the general case, the relationship between K_1', K_2', K_3' and K_4' is not restricted to a particular one. Thus, when $K_1' < K_2' < K_3' < K_4'$, positive cooperativity will be observed, and when $K_1' > K_2' > K_3' > K_4'$, negative cooperativity will be observed. More complex relationships between the K' values can also exist. Thus if $K_1' < K_2' > K_3' > K_4'$ mixed negative-positive cooperativity will be observed. The binding curve $\bar{Y}_x$ vs. [X] or $\bar{Y}_x$ log[X] will exhibit a "bump" or an intermediary plateau region (IPR). The existence of such a cooperativity was pointed out by Levitzki and Koshland (1969) and analyzed further by Teipel and Koshland (1969). They demonstrated that whenever a binding curve exhibits an intermediary plateau, it indicates that the binding protein possesses at least three ligand-binding sites. A complex kinetic response as a function of ligand saturation can also be due to changes in the catalytic constants k_1, k_2, k_3, and k_4 characterizing the catalytic turnover numbers per site in the mono, di, tri and tetrabound species (Teipel and Koshland, 1969). In this case ligand binding can be noncooperative, but the kinetic saturation curve can display cooperativity. In such a case the quantity $\frac{v}{V_{max}}$ is not identical with the saturation function $\bar{Y}_x$ as determined from direct binding studies.

Indeed, in all cases cited by Levitzki and Koshland (1969) and by Teipel and Koshland (1969), no binding measurements were reported. Rather, the behavior of the velocity versus ligand concentration displays an intermediary plateau region. This behavior can be due either to mixed negative-positive cooperativity in ligand binding or to changes in the catalytic turnover number per site as a function of ligand occupancy. It should be pointed out that only in one case, namely in yeast glyceraldehyde-3-phosphate dehydrogenase (Cook and Koshland, 1970; Mockrin et al., 1975) it was demonstrated experimentally that the binding of ligand (NAD^+) displays mixed positive-negative cooperativity. In all other cases cited from the literature by Levitzki and Koshland (1969), and by Teipel and Koshland (1969), no data are available to indicate whether the occurrence of the intermediary plateau region in the $\bar{Y}_x$ versus [ligand] plot is due to mixed cooperativity in ligand binding or to progressive changes in the catalytic constants. It is worth pointing out that in general it is found that the ratio $\dfrac{v}{v_{max}}$ obtained from kinetic measurements, represents the ligand saturation function $\bar{Y}_x$ and therefore represents the ligand saturation function expected from binding measurements.

Chapter 4
The Energy of Subunit Interactions

<u>I. Determination of Intersubunit Interaction Energy</u>

Wyman (1964) has defined the apparent subunit interaction energy for ligand binding to a protein possessing N sites according
to the equation:

$$\Delta G_I = -RT \ln \frac{K'_N}{K'_1} = -2.3\ RT \log \frac{K'_N}{K'_1} \tag{46}$$

where ΔG_I is the free energy of intersubunit interaction, K'_N the
intrinsic association constant for the Nth site, and K'_1 the intrin
sic binding constant for the first site. The apparent interaction energy is therefore defined as the amount of energy required
to change the affinity from K'_1 to K'_N. The values of K'_1 and K'_N
can be estimated from the Hill plot if enough data are accumulated at very low ligand saturation and at very high ligand saturation, namely in the regions where the Hill slope approaches
unity (Fig. 4, p. 20). The limiting slope of 1.0 at very low
concentration in fact characterizes the binding of ligand to the
first site. Therefore the crosspoint of this straight line with
a slope of 1.0 with the horizontal line of $\log \frac{\bar{Y}x}{1-\bar{Y}_x} = 0$, yields
a direct reading of $\log \frac{1}{K'_1}$ where K'_1 is the intrinsic affinity
constant. Similarly the value of $\log \frac{1}{K'_N}$ can be read graphically
from the point where the straight line with n = 1 obtained at
high ligand saturation crosses the $\log \frac{\bar{Y}x}{1-\bar{Y}_x} = 0$ line (Fig. 4,
p. 20). The free energy of intersubunit interaction can also be
read graphically directly from the Hill plot. If h is defined
as the distance between the two limiting n = 1.0 lines and l the
distance between the two values: $\log(1/K'_N)$ and $\log(1/K'_1)$ on the
log[X] axis (Fig. 4), is given by:

$$l^2 = h^2 + h^2 \tag{47}$$

or

$$l = \sqrt{2}\, h \tag{48}$$

but since

$$l = \log \frac{1}{K'_1} - \log \frac{1}{K'_N} = \log \frac{K'_N}{K'_1} \tag{49}$$

one can write, using Eq. (46):

$$\Delta G_I = -2.3RT \sqrt{2}\, h \tag{50}$$

In the case of hemoglobin, ΔG_I equals -3 kcal/mol tetramer. Since hemoglobin is a tetrahedral molecule, six intersubunit domains share this quantity of 3 kcal/mol, about 0.50 kcal per domain. It has, however, been shown that the distribution of the interaction energy is not equal among the intersubunit domains (Haber and Koshland, 1971). It is worth pointing out that strong cooperativities such as in hemoglobin (n_H = 2.8) involve low free energies. The low free energy involved in subunit interactions can either be due to hydrophobic interactions or electrostatic interactions. It is well known in hemoglobin that electrostatic salt bridges play a key role in the subunit interactions of the protein (Perutz, 1970, 1972).

Since the quantity RT at 25°C is 0.60 kcal/mol, one would expect a significant dependence of the cooperativity on temperature. This conclusion can also be reached by looking at Eq. (46), which describes the dependence of n_H on the ligand-binding constants. It is clear that the equilibrium constants characterizing the ligand-binding steps are not expected to display the same dependence on temperature. Surprisingly the effect of temperature on subunit interactions has been investigated in some detail only in a few cases (e.g., Levitzki and Koshland, 1972a).

II. The Hill Coefficient and the Intersubunit Interaction Energy

In the case of a symmetric dimer Eq. (46) obtains the form:

$$\Delta G_I = -RT \ln \frac{K'_2}{K'_1} = - 2.3 \; RT \; \log \frac{K'_2}{K'_1} \tag{51}$$

Using Eq. (32) one can obtain the relationship

$$\Delta G_I = + \frac{1}{2} \; RT \; \ln \frac{2-n_H}{n_H} = + \frac{1}{2} \; RT \; \log \frac{2-n_H}{n_H} \tag{52}$$

If one plots ΔG_I as a function of n_H, the curve obtained is quite linear over the range of $n_H = 0.4$ to 1.6 (Dahlquist and Koshland, in preparation). Thus over a significant range, the Hill coefficient is proportional to the energy of intersubunit interaction. It is clear that in the case of a tetramer K'_1 and K'_4 are not the only parameters controlling the Hill coefficient, even in a case where the binding curve is symmetric about the midpoint. This is clearly seen from Eq. (44) or (45).

III. The Meaning of Intersubunit Energy of Interaction

One can gain some intuitive insight into the meaning of subunit interaction energy by observing the schematic presentation of Figure 6.

The intersubunit interaction energy is a negative quantity in a case of positive cooperativity, and a positive quantity in negative cooperativity. This is also immediately realized from Eq. (46). In positive cooperativity, the intersubunit interaction energy is the "extra" amount of free energy which becomes available through a conformational change so that the binding improves from one binding step to the subsequent one. In negative cooperativity, the subunit interaction energy is a positive quantity characterizing the fact that conformational energy was released from the protein upon the first binding step, thus im-

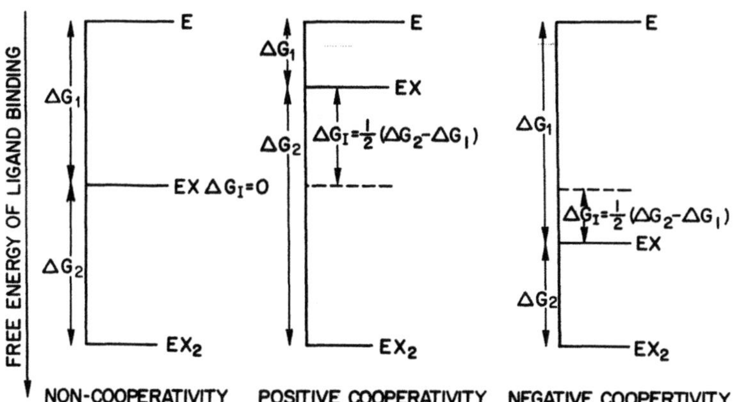

<u>Fig. 6.</u> The free energy of subunit interactions. For simplicity the dimer case is shown. The intersubunit interaction energy is the difference between the free energy of ligand binding to the second site (ΔG_2) to the free energy of binding to the first site (ΔG_1). The free energy of intersubunit interaction is given by $\Delta G_I = (\frac{\Delta G_1 + \Delta G_2}{2} - \Delta G_1)$ or $\Delta G_I = \frac{1}{2}(\Delta G_2 - \Delta G_1)$. It follows therefore that $\Delta G_I = 0$ in noncooperativity ($\Delta G_1 = \Delta G_2$), $\Delta G_I < 0$ in positive cooperativity and $\Delta G_I > 0$ in negative cooperativity. Namely, the free energy of subunit interactions is a negative quantity in positive cooperativity and a positive quantity in negative cooperativity

proving the affinity observed for that step at the expense of subsequent binding steps which now occur with diminished affinity (Fig. 6).

Chapter 5
Molecular Models for Cooperativity and Allosteric Interactions

I. Introduction

Up to this point we have discussed cooperativity only in thermodynamic terms without making any assumptions about molecular
mechanisms. It is, however, clear that deriving molecular models
for cooperativity will allow one to gain insight into the molecular events which may play a role in bringing about the large
variety of cooperative phenomena. Furthermore, defining molecular parameters which may describe cooperative phenomena may help
to design experiments to explore the mechanism of cooperativity.
Any model describing cooperativity makes use of basic experimental findings. In the development of molecular models to explain
cooperativity, hemoglobin played a key role. As early as 1935
Pauling made a serious attempt to explain the cooperative oxygen
binding to hemoglobin in terms of site-site interactions. Pauling
(1935) used a restricted Adair scheme and postulated that the
progressive increase in oxygen affinity is brought about by a
direct interaction between the heme groups. It took many years
to demonstrate that Pauling's model was incorrect, since the
heme groups are too far apart in the protein molecule to interact directly. Both the X-ray crystallographic studies of Perutz
(1970, 1972), and a variety of spectroscopic techniques revealed
that the structure of the heme groups remains unchanged, whereas
the structure of the protein is changed upon oxygen ligation.
Thus it became apparent that the cooperative nature of oxygen
binding results from interactions between protein subunits, thereby indirectly affecting the heme groups and their affinity for
oxygen. Attention was therefore focused on the nature of subunit
interactions and on their role in bringing about cooperative
phenomena. Parallel to the dramatic developments in the studies
on hemoglobin, significant progress has been made in the study

of regulatory enzymes. Monod et al. (1963, 1965) and Gerhart and Pardee (1963) showed that cooperative ligand binding is a feature of many regulatory enzymes, which were also found to possess regulatory sites (allosteric sites) regulating the function of the active site. The idea that allosteric activation or inhibition, and cooperative ligand binding can both be explained by a unified molecular theory was first put forward by Monod et al. (1965). A short time thereafter Koshland and his associates suggested an alternative molecular theory. The two molecular theories correlate the structural features of regulatory proteins with the dynamic behavior of the protein in solution. Both theories will be presented in detail, and a comparison between them will be made. It should be pointed out that for a while a rivalry existed between the two approaches, which, in fact, stimulated the understanding of regulatory phenomena. Since each group was keen on proving one of the two theories, many experimental techniques, as well as numerous diagnostic tools, were developed to explore the question: "Which of the models fits best the data". Today, more than a decade after the controversy has begun, we have reached a point where our understanding of regulatory phenomena is quite profound.

II. The Monod-Wyman-Changeux (MWC) Concerted Model

The model proposed by Monod et al. (1965) was designed to explain the well-recognized regulatory phenomena. These phenomena were: positive cooperativity, allosteric inhibition and allosteric activation. At that time phenomena such as negative cooperativity (anticooperativity) or mixed cooperativity were not recognized as such. Thus both the original MWC concerted model as well as the original KNF (Koshland-Némethy-Filmer) sequential model were derived to explain positive cooperativity, allosteric activation and allosteric inhibition.

1. Basic Assumptions of the Concerted Model

1. Allosteric proteins are oligomers composed of protomers (sub-units) associated in such a way that they all occupy equivalent positions within the oligomeric molecule. This geometric arrangement implies that the molecule possesses at least one axis of symmetry.

2. All protomers possess identical stereospecific binding sites for the ligands which bind to the oligomeric protein. The symmetry relationship between each set of stereospecific binding sites is identical to the symmetry of the molecule.

3. The ligand-free enzyme exists in two conformations T (taught) and R (relaxed), which are in equilibrium. These states differ in the free energy of interaction between the subunits. Thus they differ in the conformational constraints imposed on the promoters. These differences are reflected in both the tertiary and the quarternary structure of the protein.

4. The two (or more) conformational states differ in their affinity towards each set of ligands.

5. When the protein undergoes a change from one state to another state, its molecular symmetry is conserved.

We shall analyze the Monod-Wyman-Changeux model first for the allosteric dimer and then for the case of an N-mer. We have seen in previous sections that the allosteric dimer can exhibit many of the allosteric phenomena known, and that the equations governing its behavior are rather simple and readily derived. We have also seen that exact analytical expressions can be obtained for the dimer case with no assumptions made. Thus the analysis of a dimer is very beneficial in terms of understanding the basic allosteric phenomena.

2. The Allosteric Dimer Analyzed by the MWC Model

We shall now consider two cases: the case of exclusive binding and the case of nonexclusive binding to a protein dimer (Fig. 7).

Exclusive Binding. In this case the binding of ligand occurs to the R-state only (Fig. 7):

$$T \xrightleftharpoons{L} R \xrightleftharpoons{K'_R} RX \xrightleftharpoons{K'_R} RX_2 \tag{53}$$

Models:

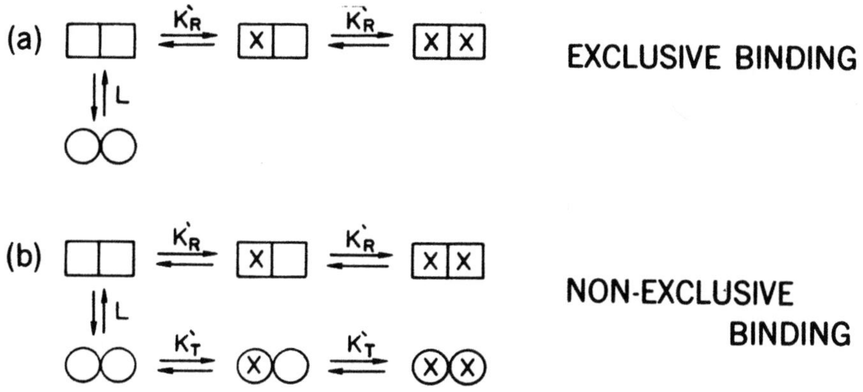

(a) [ ][ ] $\xrightleftharpoons{K_R'}$ [x][ ] $\xrightleftharpoons{K_R'}$ [x][x]   **EXCLUSIVE BINDING**

$\updownarrow L$

◯◯

(b) [ ][ ] $\xrightleftharpoons{K_R'}$ [x][ ] $\xrightleftharpoons{K_R'}$ [x][x]   **NON-EXCLUSIVE BINDING**

$\updownarrow L$

◯◯ $\xrightleftharpoons{K_T'}$ (x)◯ $\xrightleftharpoons{K_T'}$ (x)(x)

Mathematics:

Case (a)

$$K_1' = \frac{K_R'}{1+L}$$

$$K_2' = K_R'$$

$$n_H = \frac{2}{1+\sqrt{\dfrac{1}{1+L}}}$$

Case (b)

$$K_1' = \frac{K_T'L + K_R'}{1+L}$$

$$K_2' = \frac{K_T'^2 L + K_R'^2}{K_T'L + K_R'}$$

$$n_H = \frac{2}{1+\sqrt{\dfrac{(K_T'L + K_R')^2}{(1+L)(K_T'^2 L + K_T'^2)}}}$$

Definitions:

$$L = \frac{[\,◯◯\,]}{[\,\square\,]}$$

$$K_R' = \frac{[\,\boxed{x}\,]}{[\,\square\,][x]}$$

$$K_T' = \frac{[\,⊗\,]}{[\,◯\,][x]}$$

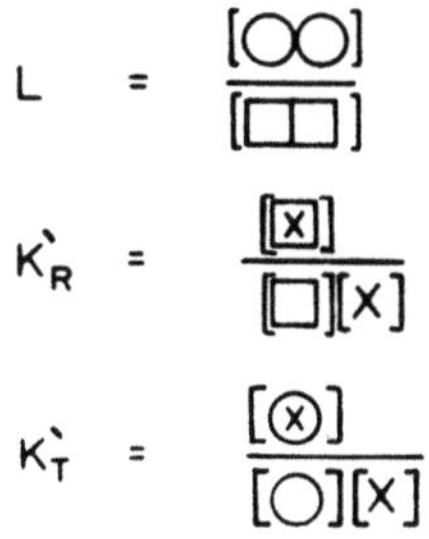

<u>Fig. 7.</u> The Monod-Wyman-Changeux model for an allosteric dimer

L is the allosteric equilibrium constant:

$$L = \frac{[T]}{[R]} \tag{54}$$

where [T] is the concentration of unliganded protein in the T-state and [R] is the concentration of unliganded protein in the R-state.

K'_R is the intrinsic ligand affinity (association constant) to the R-state. In this case, if [R] is expressed in terms of [E], the free enzyme concentration, it is clear that:

$$[E] = [T] + [R] \tag{55}$$

using Eqs. (54) and (55) one obtains:

$$[E] = [R]L + [R] \tag{56}$$

or

$$[R] = \frac{[E]}{1 + L} \tag{57}$$

Thus:

$$[RX] = 2K'_R \frac{1}{1 + L} [E][X] \tag{58}$$

and

$$[RX_2] = K'^2_R \frac{1}{1 + L} [E][X]^2 \tag{59}$$

Applying Eq. (29) for the symmetric dimer ($K_1 = K'_1$ and $K_2 = K'_2$) one obtains:

$$[RX] = 2K'_1 [R][X] \tag{60}$$

$$[RX_2] = K'_1 K'_2 [R][X]^2 \tag{61}$$

where K'_1 and K'_2 are the intrinsic affinity constants derived from a model independent thermodynamic treatment (Fig. 5).

Equating Eqs. (60) and (61) with the Eqs. (58) and (59), one obtains the relationships:

$$K_1' = \frac{K_R'}{1 + L} \tag{62}$$

and

$$K_2' = K_R' \tag{63}$$

Inserting Eqs. (62) and (63) into the general expression for the Hill coefficient at 50% for the symmetric dimer [Eq. (32)] one obtains:

$$n_H = \frac{2}{1 + \sqrt{\dfrac{1}{1 + L}}} \tag{64}$$

From this relationship it follows that $1.0 \leq n_H \leq 2.0$, since $L \geq 0$. Thus the MWC model for the exclusive binding case cannot account for negative cooperativity where n_H is found to be less than 1.0. It should be stressed that Eq. (32), which is model-independent, allows for negative cooperativity.

Nonexclusive Binding. In the nonexclusive case, both the T-form and the R-form bind the ligand with intrinsic binding constants K_T' and K_R', respectively, where $K_T' < K_R'$ (Fig. 7). In this case one can write:

$$[EX] = [TX] + [RX] = 2K_T'[T][X] + 2K_R'[R][X] \tag{65}$$

and:

$$[EX] = 2\left(\frac{K_T'L}{1 + L} + \frac{K_R'}{1 + L}\right)[E][X] = 2\left(\frac{K_T'L + K_R'}{1 + L}\right)[E][X] \tag{66}$$

Therefore

$$[EX_2] = [TX_2] + [RX_2] = \frac{K'^2_T L + K'^2_R}{1 + L} [E][X]^2 \qquad (67)$$

Equating Eqs. (66) and (67) with the Eqs. (60) and (61) yields the following relationships:

$$K'_1 = \frac{K'_T L + K'_R}{1 + L} \qquad (68)$$

$$K'_2 = \frac{K'^2_T L + K'^2_R}{K'_T L + K'_R} \qquad (69)$$

In this case the expression for the Hill coefficient at 50% for a symmetric dimer assumes the form:

$$n_H = \frac{2}{1 + \sqrt{\dfrac{(K'_T L + K'_R)^2}{(1 + L)(K'^2_T L + K'^2_R)}}} \qquad (70)$$

One can see that when $L = 0$, $n_H = 1.0$, and when $L > 0$, $n_H > 1$. As in the case of exclusive binding: $1.0 \leq n_H \leq 2.0$. Again, negative cooperativity cannot be accounted for by the concerted model of Monod et al.

3. Allosteric Inhibition and Allosteric Activation in the MWC Model

The Monod-Wyman-Changeux model offers an elegant molecular explanation for allosteric inhibition and allosteric activation. The allosteric inhibitor I has a higher affinity for the T-state, whereas the allosteric activator A has a higher affinity for the R-state. Therefore, in the presence of the allosteric inhibition, a larger fraction of the enzyme is pulled to the T-state, thereby increasing the apparent value of L. The apparent affinity of the protein towards the substrate, X, is thus reduced and the response to X becomes more-cooperative. Similarly, in the presence of the allosteric activator, the apparent affinity of the system towards the ligand is increased, and the curve becomes less cooperative towards the substrate (Fig. 8). Upon saturation of the system with the activator, the system may become Michaelian towards the substrate if the R-state binds the activator exclusively (Fig. 8).

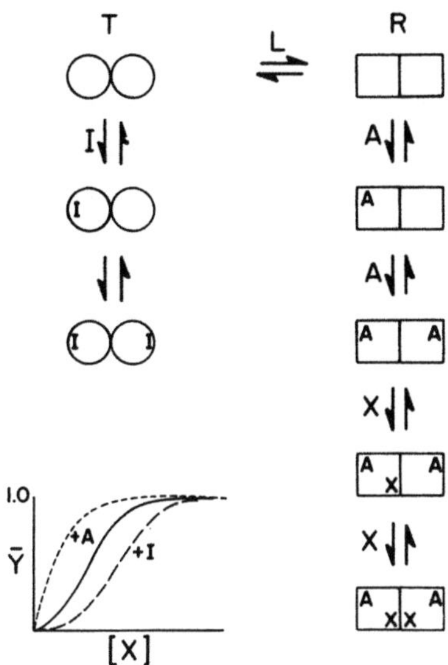

Fig. 8. The effect of an allosteric activator and of an allosteric inhibitor according to the MWC model. In this figure a simplified scheme for the effect of an activator A and of an inhibitor I on the binding curve for the substrate X is used. It is assumed that the activator binds exclusively to the R-state, the inhibitor I binds exclusively to the T-state and the substrate X binds exclusively to the R-state. The effect of I and A on the shape and the midpoint of the saturation curve is schematically depicted at the *left corner* of the figure. The binding of X in the presence of A is Michaelian (hyperbolic) with higher affinity (lower $[X_{0.5}]$) as compared with the saturation curve in the absence of A. The reverse situation exists in the presence of I. Thus in the presence of I the binding curve is more cooperative and occurs with lower affinity (higher $[X_{0.5}]$). It is clear that the scheme may be more complex where all ligands bind to both the R- and the T-states. Under these conditions the effects of A and I will be less dramatic

4. The General Case

Let us consider an enzyme with N subunits which occurs at two conformations T and R in equilibrium with each other. Ligand binding can occur to both the R-state and the T-state. The ma-

40

thematical treatment follows the original publications of Monod
et al. (1965) and of Rubin and Changeux (1966). In the absence
of ligand, the two states symbolized as Ro and To are in equilibrium where L is the allosteric equilibrium constant (allosteric constant):

$$R_O \rightleftharpoons T_O \tag{71}$$

$$L = \frac{[T_O]}{[R_O]} \tag{72}$$

Let us consider the binding of a ligand X to the two forms of
the protein: to the high affinity from R, and to the low affinity form T. Since each of the forms is symmetric, the affinity of
the ligand to each of the forms is describable by a single association constant K_R' (to the R-form) and K_T' (to the T-form):

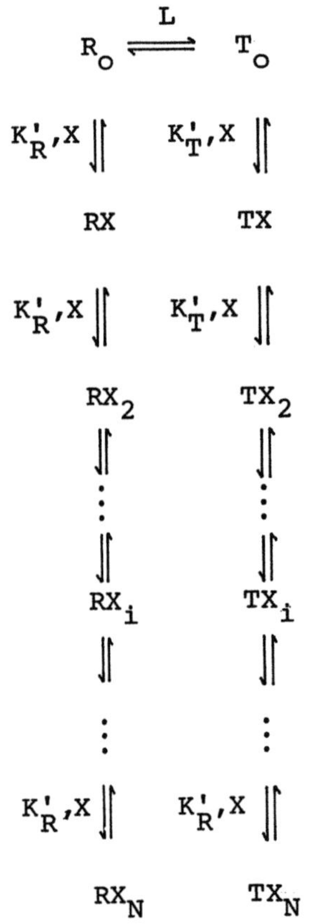

$$
\begin{array}{ccc}
 & L & \\
R_O & \rightleftharpoons & T_O \\
K_R',X \updownarrow & & K_T',X \updownarrow \\
RX & & TX \\
K_R',X \updownarrow & & K_T',X \updownarrow \\
RX_2 & & TX_2 \\
\updownarrow & & \updownarrow \\
\vdots & & \vdots \\
\updownarrow & & \updownarrow \\
RX_i & & TX_i \\
\updownarrow & & \updownarrow \\
\vdots & & \vdots \\
K_R',X \updownarrow & & K_R',X \updownarrow \\
RX_N & & TX_N
\end{array}
$$

The fraction of sites actually bound with ligand is given by:

$$\bar{Y}_x = \frac{1}{N} \times \frac{[RX] + 2[RX_2] + \ldots + N[RX_N] + [TX] + 2[TX_2] + \ldots + N[TX_N]}{[R] + [RX] + [RX_2] + \ldots + [RX_N] + [T] + [TX] + [TX_2] + \ldots + [TX_N]}$$

$$(73)$$

It is clear that:

$$[RX] = NK_R'[R][X] \qquad\qquad [TX] = NK_T'[T][X]$$

$$[RX_2] = \frac{N-1}{2} K_R'[RX][X] \qquad [TX_2] = \frac{N-1}{2} K_T'[TX][X]$$

$$\vdots \qquad\qquad\qquad\qquad \vdots$$

$$[RX_N] = \frac{1}{N} \times K_R'[RX_{N-1}][X] \qquad [TX_N] = \frac{1}{N} \times K_T'[TX_{N-1}][X]$$

Following Monod et al. (1965), one can define the following parameters:

$$\alpha = K_R'[X] = \frac{[X]}{K_R}$$

$$(74)$$

K_R' is the intrinsic association constant, and K_R is the intrinsic dissociation constant to the R-state respectively.

$$c = \frac{K_T'}{K_R'} = \frac{K_R}{K_T} \qquad \text{or} \qquad K_T' = cK_R'$$

$$(75)$$

where K_T' is the intrinsic association constant and K_T is the intrinsic dissociation constant to the T-state respectively. Using these relationships one can show that Eq. (73) obtains the form:

$$\bar{Y}_x = \frac{Lc\alpha(1 + \alpha)^{N-1} + \alpha(1 + \alpha)^{N-1}}{L(1 + c\alpha)^N + (1 + \alpha)^N}$$

$$(76)$$

Thus the saturation curve can be described by Eq. (76) using four parameters: L, c, K_R', and N. N is the number of total binding sites per oligomer and can, in principle, be determined directly. However, in most cases the exact number of binding sites per oligomer is obtained from ligand-binding measurements fitted to Eq. (76). Thus in fact $\bar{Y}_x$ is a function of four parameters (including N).

Using the definitions (74) and (75), Eq. (76) can be written also as:

$$\bar{Y}_x = \frac{LK_T'[X](1+K_T'[X])^{N-1} + K_R'[X](1+K_R'[X])^{N-1}}{L(1+K_T'[X])^N + (1+K_R'[X])^N} \qquad (77)$$

It can be seen that when $K_T' = K_R'$ (c = 1) or when L = 0, Eq. (75) simplifies to:

$$\bar{Y}_x = \frac{\alpha}{1 + \alpha} \qquad (78)$$

or

$$\bar{Y}_x = \frac{K_R'[X]}{1 + K_R'[X]} = \frac{[X]}{K_R + [X]} \qquad (79)$$

Equations (78) and (79) describe in fact the classical noncooperative Michaelis-Henri case. The model therefore accounts well for positive homotropic (positively cooperative) binding where the limiting case is the noncooperative binding case described by the Michaelis-Henri equation. Equation (76) describes ligand binding in the nonexclusive case, where the ligand has a finite affinity to both states of the protein: R and T. In the exclusive binding case where $K_T' = 0$ (c = 0) Eq. (76) simplifies to:

$$\bar{Y}_x = \frac{\alpha(1 + \alpha)^{N-1}}{L + (1 + \alpha)^N} \qquad (80)$$

and Eq. (77) simplifies to:

$$\bar{Y}_x = \frac{K_R'[X](1 + K_R'[X])^{N-1}}{L + (1 + K_R'[X])^N} \qquad (81)$$

Heterotropic Effects. According to the model, heterotropic effects are due exclusively to the displacement of the R to T equilibrium by the regulatory ligands. An allosteric inhibitor I will bind to the T-state with higher affinity than to the R-state, whereas the allosteric activator, A, will bind to the R-state with higher affinity than to the T-state. In this fashion heterotropic ligands affect the homotropic interactions of the ligand X, as is in fact observed experimentally. Thus the ligands I and A affect the allosteric constant L, and the apparent allosteric constant L' becomes a function of the concentrations of I and A and their respective affinity constants to the T-state and the R-state. The correlation between L' and L is given by the expression:

$$L' = \frac{\sum_0^N T_I}{\sum_0^N R_A} L \qquad (82)$$

where $\sum_0^N T_I$ and $\sum_0^N R_A$ stand respectively for the sum of the different complexes of the T-state with I, and of the R-state with A. In the case which we are discussing here, it is assumed that I possesses negligible affinity to the R-state, and A possesses negligible affinity to the T-state. If I binds to the T-state with an affinity of K_I', and A binds with an affinity of K_A' to the R-state, Eq. (82) will obtain the form:

$$L' = L \frac{(1 + K_I'[I])^N}{(1 + K_A'[A])^N} \qquad (83)$$

If one defines:

$$\beta = K_I'[I] \quad \text{and:} \quad \gamma = K_A'[A]$$

Eq. (83) will obtain the form:

$$L' = L\left(\frac{1 + \beta}{1 + \gamma}\right)^N \tag{84}$$

and Eq. (76) will obtain the form:

$$\bar{Y}_x = \frac{L'c\alpha(1 + c\alpha)^{N-1} + \alpha(1 + \alpha)^{N-1}}{L'(1 + c\alpha)^N + (1 + \alpha)^N} \tag{85}$$

and Eq. (77) will obtain the form:

$$\bar{Y}_x = \frac{L'K_T'(1+K_T')[X])^{N-1} + K_R'[X](1+K_R'[X])^{N-1}}{L'(1+K_T'[X])^N + (1+K_R'[X])^N} \tag{86}$$

The model can be extended to a case where I and A bind to both the R-state and the T-state. This case was discussed in some detail by Rubin and Changeux (1966). The value of L' can be determined by fitting of ligand-binding data to the general Eq. (85) or (86). A simpler method to determine L' was pointed out by Rubin and Changeux (1966). The method involves the measurement of $\alpha_{\frac{1}{2}}$, namely the substrate concentration yielding 50% saturation (at $\bar{Y}_x = \frac{1}{2}$). Under these conditions one obtains a correlation between L' and $\alpha_{\frac{1}{2}}$:

$$L' = \frac{\alpha_{\frac{1}{2}} - 1}{1 - \alpha_{\frac{1}{2}}c}\left(\frac{1 + \alpha_{\frac{1}{2}}}{1 + \alpha_{\frac{1}{2}}c}\right)^{N-1} \tag{87}$$

This relationship is derived from Eq. (85) by inserting $\bar{Y}_x = \frac{1}{2}$. Equation (87) predicts that the correlation between $\log \alpha_{\frac{1}{2}}$ and L' will be sigmoidal.

5. *Phenomena Explained by the Concerted Model*

A broad spectrum of phenomena can be accounted for by the Monod-Wyman-Changeux allosteric model. These phenomena include:
1. Positively cooperative ligand binding.
2. Allosteric inhibition and allosteric activation.
3. Incomplete allosteric inhibition or activation are easily accommodated by the more general nonexclusive binding case.
4. Changes in k_{cat} or in $[X_{0.5}]$ or in both in the response of an enzyme towards its substrate, induced by allosteric ligands can be explained. It is easy to visualize that the R-state and the T-state differ not only in their ligand affinity, but also in their respective catalytic efficiency (k_{cat}).

6. *Cooperativity in the Monod-Wyman-Changeux Model*

In this section an attempt will be made to correlate the most commonly used index for cooperativity: the Hill coefficient at 50% saturation, n_H, with the molecular parameters of the Monod-Wyman-Changeux concerted model.

1. The Dimer Case. We have seen earlier that the Hill coefficient can be easily described by the parameters of the MWC model in the case of a dimer [Eqs. (64) and (70), Fig. 7].

2. The General Case. The MWC model is capable of describing the binding of ligand to an oligomeric protein [Eq. (76)] when both the T and R forms of the enzyme can bind the ligand. In the exclusive binding case, the binding of ligand is given by Eq. (80).

It is interesting to correlate the MWC paramters with n_H, which is a model independent quantity, and can always be measured. The discussion below is based on the analysis performed by Dahlquist and Koshland (in preparation). We shall consider both the exclusive binding case and the nonexclusive binding case.

a) Exclusive Binding. It can be shown (see Appendix) that the midpoint concentrations of ligand, $[X_{0.5}]$, are related to the MWC parameters K'_R and L in the exclusive binding case (no binding to the T-state) by a fairly simple relationship:

46

$$L = (K_R'[X_{0.5}] + 1)^{N-1} (K_R'[X_{0.5}]-1) \tag{88}$$

One can also prove that the Hill coefficient, n_H, is related to K_R' and $[X_{0.5}]$ by the following formula:

$$K_R'[X_{0.5}] = \frac{N + n_H - 2}{N - n_N} \tag{89}$$

The full derivation of Eqs. (88) and (89) is given in the Appendix. Equations (88) and (89) can be used to evaluate L and K_R' from binding data without resorting to curve-fitting procedures.

From Eqs. (89) and (88) it is apparent that the Hill coefficient is dependent only on L. This result is intuitively well understood, since in the absence of ligand binding to the T-state ($K_T' = 0$), L alone will determine the shape, and therefore the cooperativity of the binding curve.

b) *Nonexclusive Binding.* In the general case where both the R-state and the T-state bind the ligand, the dependence of the Hill coefficient on the parameters of the MWC model becomes more complex. For a fixed value of L it is clear that if one allows ligand binding to the T-state [c > 0 in Eq. (75)], the Hill coefficient will decrease. One can evaluate both the midpoint concentration yielding 50% saturation ($\bar{Y}_x = \frac{1}{2}$), and the Hill slope at that point (n_H). From Eq. (76) it is apparent that the measured parameters $[X_{0.5}]$ and n_H are related to three unknowns: L, c and K_R'. Thus n_H depends on both c and L. It is clear that at a fixed L value the cooperativity will decrease as the value of c increases.

From Eq. (76) one can derive an expression for L at the midpoint ($\bar{Y}_x = \frac{1}{2}$):

$$L = \frac{(1 + \alpha_{0.5})^{N-1} (1 - \alpha_{0.5})}{(1 + c\alpha_{0.5})^{N-1} (c\alpha_{0.5} - 1)} \tag{90}$$

From Eqs. (23) and (76) one can derive an expression for the Hill coefficient, n_H, at the midpoint:

$$n_H = 1 + \frac{(N-1)L(1-c)^2\alpha(1+\alpha)^{N-2}(1+c\alpha)^{N-2}}{[(1+\alpha)^{N-1}+L\alpha(1+c\alpha)^{N-1}][(1+\alpha)^{N-1}+L(1+c\alpha)^{N-1}]} \tag{91}$$

The Hill coefficient can assume values equal to 1.0 (when $L = 0$) or greater than 1.0 (when $L > 0$), but never values less than 1.0. Thus in no case can the MWC model account for negative cooperativity, but can account for almost every case of positive cooperativity. Using Eq. (91) one can show that n_H will reach a maximum for fixed c values, as was pointed out by Rubin and Changeux (1966).

III. The Koshland-Némethy-Filmer (KNF) Sequential Model

The model of Koshland, Némethy and Filmer proposed in 1966 examines in detail the role of the subunit interfaces in determining the regulatory behavior of the protein. Unlike the Monod-Wyman-Changeux concerted model the existence of hybrid conformational states in which protein symmetry is not conserved is allowed.

1. Basic Assumptions of the Sequential Model

The fundamental postulates of the model can be summarized as follows:
1. The oligomeric protein exists in one conformation prior to ligand binding.
2. Upon ligand binding a conformational change is induced in the subunit binding the ligand.
3. The conformational change is not restricted to the filled subunit, but may be transmitted to neighboring vacant subunits via the intersubunit domains.
4. The ligand binds preferentially to one of the conformations. As with the MWC model, one may most easily understand the KNF sequential model by looking at the case of an allosteric dimer.

48

2. *The Allosteric Dimer Analyzed by the KNF Model*

In Figure 9 the scheme of ligand binding to a dimer according to the KNF sequential model is given. Definitions of the different parameters are also given in the figure: Kt_{AB} is the equilibrium constant of the square conformation (Conformation B) over the round one (Conformation A). K_{X_B} is the intrinsic affinity constant of the B conformation towards X. K_{AB} is the subunit interaction constant between a circle (A) conformation and a square (B) conformation divided by K_{AA}. K_{BB} is defined as the subunit interaction between two B conformations divided by K_{AA} (Fig. 9).

It is easily seen from Figure 9 how the Adair equation is derived for the KNF model, and how the expression for the Hill coefficient n_H at the midpoint is arrived at:

$$n_H = \frac{2}{1 + \sqrt{\dfrac{K_{AB}^2}{K_{BB}}}} \; ; \; [X_{0.5}] = \psi_2^{-\frac{1}{2}} = (K_1' K_2')^{-\frac{1}{2}} = K_{BB}^{-\frac{1}{2}} (K_{X_B} K_{t_{AB}})^{-1} \tag{92}$$

It can be seen from Eq. (92) that the ratio between K_{AB}^2 and K_{BB} will determine whether n_H will be larger than, smaller than, or equal to 1.0. It is interesting that when $K_{AB}^2 = K_{BB}$ then $n_H = 1.0$; this means that not every conformational change coupled to ligand binding in a multisubunit enzyme must result in cooperativity. On the contrary, conformational changes which obey the relation $K_{AB}^2 = K_{BB}$ will result in noncooperative saturation curves. Indeed, in many multisubunit enzymes the ligand-binding curves are noncooperative, although conformational changes do occur. Thus, for example, in the tetrameric enzyme lactate dehydrogenase, large conformational changes accompany NAD^+ binding, but the binding of the latter is noncooperative. The conformational changes which occur within the lactate dehydrogenase tetramer were observed by Rossmann and his colleagues using X-ray crystallographic techniques as well as by other techniques (Rossmann et al., 1971; Evars and Kaplan, 1973). It is immediately seen from Eq. (92) that negative cooperativity will occur when $K_{AB}^2 > K_{BB}$. Thus, the sequential model can easily account for nega-

Models:

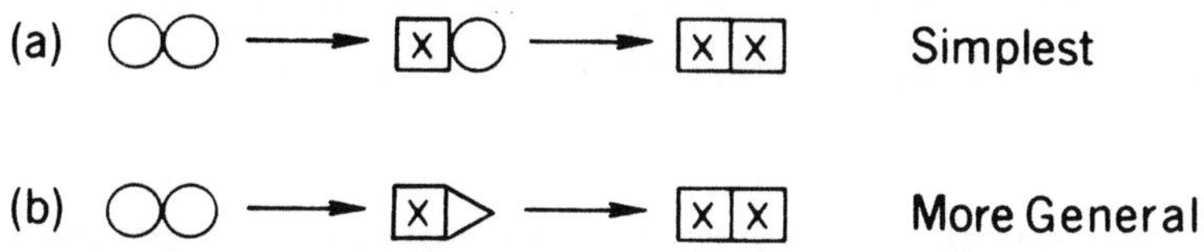

(a) ⟨two circles⟩ ⟶ ⟨box-x + circle⟩ ⟶ ⟨box-x box-x⟩   Simplest

(b) ⟨two circles⟩ ⟶ ⟨box-x + triangle⟩ ⟶ ⟨box-x box-x⟩   More General

---

Mathematics:

Case (a)

$$K_1' = K_{t_{AB}} K_{x_B} K_{AB}$$

$$K_2' = K_{t_{AB}} K_{x_B} \frac{K_{BB}}{K_{AB}}$$

$$n_H = \frac{2}{1 + \sqrt{\dfrac{K_{AB}^2}{K_{BB}}}}$$

Case (b)

$$K_1' = K_{t_{AB}} K_{t_{AC}} K_{x_B} K_{BC}$$

$$K_2' = K_{t_{CB}} K_{x_B} \frac{K_{BB}}{K_{BC}}$$

$$n_H = \frac{2}{1 + \sqrt{\dfrac{K_{t_{AB}} K_{t_{AC}}}{K_{t_{CB}}} \times \dfrac{K_{BC}^2}{K_{BB}}}}$$

Definitions:

$$K_{t_{AB}} = \frac{[\square]}{[\bigcirc]} \qquad K_{t_{AB}} = \frac{[\square]}{[\bigcirc]} \qquad K_{AB} = \frac{[\bigcirc\square][\bigcirc]}{[\bigcirc\bigcirc][\square]}$$

$$K_{x_B} = \frac{[\boxtimes]}{[\square]} \qquad K_{t_{AC}} = \frac{[\triangleright]}{[\bigcirc]} \qquad K_{CB} = \frac{[\square\triangleright][\bigcirc]^2}{[\bigcirc\bigcirc][\square][\triangleright]}$$

$$K_{AB} = \frac{[\bigcirc\square][\bigcirc]}{[\bigcirc\bigcirc][\square]} \qquad K_{t_{CB}} = \frac{[\square]}{[\triangleright]} \qquad K_{BB} = \frac{[\square\square][\bigcirc]^2}{[\bigcirc\bigcirc][\square]^2}$$

$$K_{BB} = \frac{[\square\square][\bigcirc]^2}{[\bigcirc\bigcirc][\square]^2} \qquad K_{x_B} = \frac{[\boxtimes]}{[\square][x]}$$

$$K_{AA} = \frac{[\bigcirc\bigcirc]}{[\bigcirc]^2} \equiv 1.0$$

Fig. 9. The general KNF sequential model for an allosteric dimer. Two cases (a and b) are considered. In both cases the protein in the free state is a symmetric structure. In case *a* the subunit assumes a total of two different conformations and in case *b* the subunit can assume a total of three different confirmations

tive cooperativity. It is also seen from Eq. (92) that positive cooperativity will occur when $K_{BB} > K_{AB}^2$. The KNF model can therefore account for all types of cooperativity predicted by the Adair approach.

Figure 9 depicts two cases for an allosteric dimer, which obey the KNF sequential model. In the simple case, only two conformations of the subunit are involved, whereas in the more complex case the subunit can assume three conformations. It is possible to extend the sequential model to its most general scheme as represented in Figure 9.

The cases considered in Figure 9 yield more complex algebraic expressions. It is usually found, in real cases, that the scheme as presented in case (a) of Figure 9 is sufficient to account for the experimental data. This state of affairs does not mean that the protein subunit cannot assume more than two conformations, but rather that the quality of the binding data is such that binding equations with a minimal number of parameters are usually sufficient to account for the binding phenomena. In fact it is known from physicochemical and chemical studies of regulatory proteins that the protein subunit can assume a number of conformations (e.g., Schlessinger and Levitzki, 1974; Levitzki and Koshland, 1972b).

3. *Allosteric Activation and Allosteric Inhibition in the KNF Model – the Dimer Case*

Let us consider the general case, where the protein can bind the ligand X, the allosteric activator A, and the allosteric inhibitor I, and treat it according to the KNF sequential model (Fig. 10). It can be seen from Figure 10 that a number of assumptions have already been made implicitly. The first is that the square conformation is the final subunit conformation whether the activator A is bound or the substrate X is bound. It can easily be visualized that a subunit with A bound can attain still a different conformation, depending on whether X is bound to it or to its neighboring subunit. The second assumption is that the enzyme does not bind A and I simultaneously, and that a subunit to which I is bound is incapable of binding X.

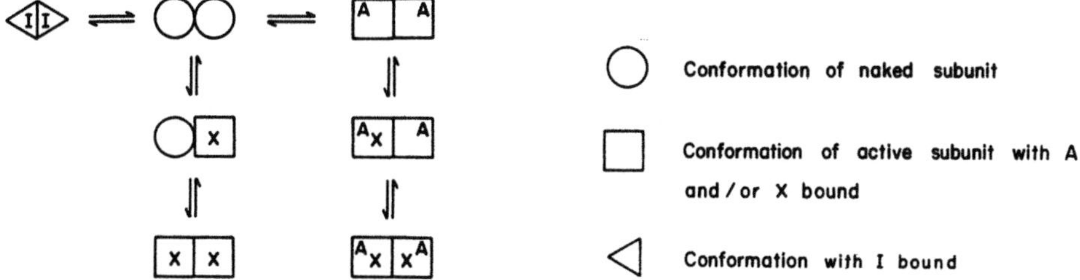

Fig. 10. Allosteric activation and allosteric inhibition according to the sequential model. For simplicity it is assumed that the subunit can assume three different conformations: the one typical for the unbound state, the one typical of the inhibitor (I) bound state and the one typical for the activator (A) bound state. It is also assumed that the conformation of the substrate (X) bound state is identical to that of the activator bound state. It is also assumed that the inhibitor bound state is incapable of binding X or A. This kind of a scheme is usually sufficient to explain the increase in the cooperativity towards X and the decrease in the affinity towards X in the presence of I. This scheme is also sufficient to explain the noncooperativity in X binding in the presence of A and the increase in the affinity towards X in the presence of A

The scheme in Figure 10 shows that many terms will appear in the Adair equation, each one specifying the specific combination of conformations. There is no difficulty in fitting binding curves by adjusting many parameters as in the case discussed here. The major problem is to find ways of proving the existence of the different conformations, or the absence of some conformations. This becomes a formidable task, even in the case of a dimer. One must explore the existence of different conformations of the protein subunit by direct physicochemical and chemical methods. As a guideline, therefore, one should fit binding curves with the maximal number of conformations that can be accounted for experimentally. As in the case of the concerted model, one usually finds it difficult to prove the existence of more than three conformations. Only in a limited number of cases has direct evidence for a multitude of conformational states been obtained (Schlessinger and Levitzki, 1974; Levitzki and Koshland, 1972b).

4. The KNF Model – the Tetramer Case

In contrast to the MWC model, the KNF model defines special parameters for the subunit interaction across intersubunit interfaces. Therefore, the type of intersubunit interface depends on

the geometric architecture of the multisubunit protein. In the
case of a dimer, only one interface is involved, whereas in more
complex structures the type of interface to be considered depends
on the architecture of the molecule.

Therefore, the KNF model, unlike the MWC model, cannot yield a
general and at the same time a relatively simple equation such
as Eqs. (76) and (77) of the MWC model. We shall therefore consider in this section the case of a tetrameric protein following
almost without change the original formulations of Koshland et al.
(1966). Since many, if not most, regulatory proteins are tetrameric, it is worthwhile to analyze the tetramer case.

In the analysis given below, it is assumed that all the subunit
interactions are identical in all directions in space. Two cases
will be considered: the square model and the tetrahedral model.
In the former case each subunit makes two identical contacts
with its neighbors, whereas in the latter three identical contacts are made between each subunit and its neighbors. In Figure
11 a schematic representation of ligand binding to the "square"
molecule and the "tetrahedral" molecule is shown.

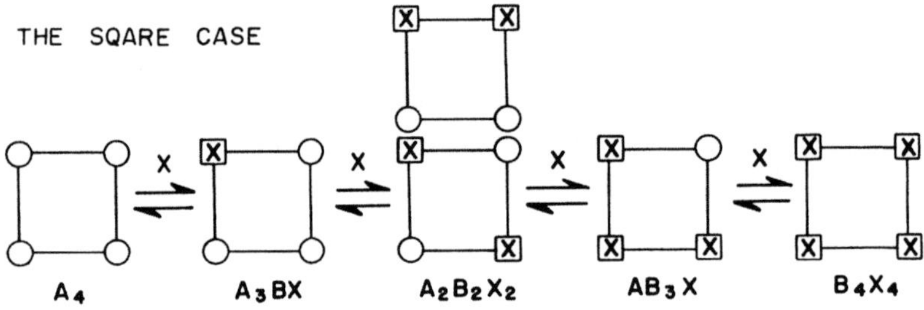

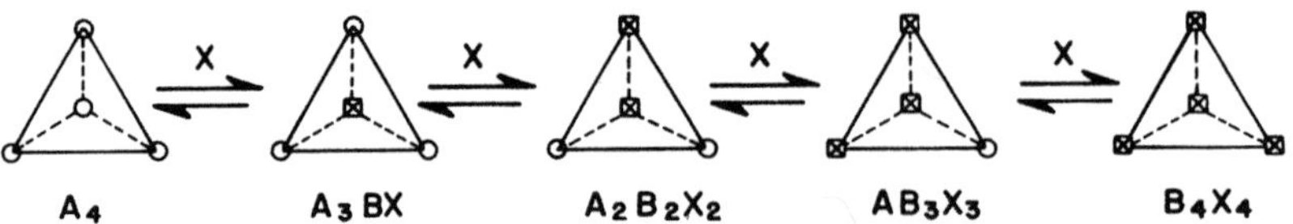

Fig. 11. The binding of a ligand to a tetramer according to the KNF model.
A is the subunit conformation in the absence of ligand (*circle*) and B is the
subunit conformation in the presence of ligand (*rectangular*)

1. The "Square" Case. In the square case the ligand-bound species are:

$$[EX] = \psi_1 [A_4][X] = [A_3BX] = 4K_{AB}^2 K_{t_{AB}}[A_4][X] \tag{93}$$

$$[EX_2] = \psi_2[A_4][X]^2 = [A_2B_2X_2] = (4K_{AB}^2 K_{BB} + 2K_{AB}^4)[K_{x_B} K_{t_{AB}}]^2 [A_4][X]^2 \tag{94}$$

$$[EX_3] = \psi_3[A_4][X]^3 = [AB_3X_3] = 4K_{AB}^2 K_{BB}^2 (K_{x_B} K_{t_{AB}})^3 [A_4][X]^3 \tag{95}$$

$$[EX_4] = \psi_4[A_4][X]^4 = K_{BB}^4 (K_{x_B} K_{t_{AB}})^4 [A_4][X]^4 \tag{96}$$

The corresponding values of the intrinsic assoication constants K_i' are obtained by taking the ratio of ψ_i/ψ_{i-1} and using the appropriate statistical factor as shown in Table 1.

Table 1. The values ψ_1, ψ_2, ψ_3 and ψ_4 according to the simplified KNF square and tetrahedral models

	Square	Tetrahedral
ψ_1	$4K_{AB}^2 (K_{x_B} K_{t_{AB}})$	$4K_{AB}^3 (K_{x_B} K_{t_{AB}})$
ψ_2	$2K_{AB}^4 + 4K_{BB}K_{AB}^2 (K_{x_B} K_{t_{AB}})^2$	$6K_{AB}^4 K_{BB} (K_{x_B} K_{t_{AB}})^2$
ψ_3	$4K_{AB}^2 K_{BB}^2 (K_{x_B} K_{t_{AB}})^3$	$4K_{AB}^3 K_{BB}^3 (K_{x_B} K_{t_{AB}})^3$
ψ_4	$K_{BB}^4 (K_{x_B} K_{t_{AB}})^4$	$K_{BB}^6 (K_{x_B} K_{t_{AB}})^4$
Does the equality $\psi_1^2\psi_4 = \psi_3$ hold?	yes	yes

The values ψ_1 through ψ_4 are related to the intrinsic affinity constants and to the macroscopic thermodynamic binding constants (the Adair constants) by the following relationships (see also Ch. 3.III):

$$\psi_1 = K_1 = 4K_1'; \quad \psi_2 = K_1K_2 = 6K_1'K_2'; \quad \psi_3 = K_1K_2K_3 = 4K_1'K_2'K_3'; \quad \text{and}$$
$$\psi_4 = K_1K_2K_3K_4 = K_1'K_2'K_3'K_4'$$

From these relationships one can calculate the expressions for the thermodynamic association constants and the intrinsic binding affinities in terms of the KNF model.

The binding equation for the square case is given by Eq. (97):

$$N_x = \frac{4K_{AB}^2(K_{x_B}K_{t_{AB}})[X]+4(K_{AB}^4+2K_{AB}^2K_{BB})K_{x_B}K_{t_{AB}})^2[X]^2}{1+4K_{AB}^2(K_{x_B}K_{t_{AB}})[X]+2(K_{AB}^4+2K_{AB}^2K_{BB})K_{x_B}K_{t_{AB}}[X]^2}$$

$$\frac{+12K_{AB}^2K_{BB}^2(K_{x_B}K_{t_{AB}})^3[X]^3+4K_{BB}^4(K_{x_B}K_{t_{AB}})^4[X]^4}{+4K_{AB}^2K_{BB}^2(K_{x_B}K_{t_{AB}})^3[X]^3+K_{BB}^4(K_{x_B}K_{t_{AB}})^4[X]^4} \tag{97}$$

and the fractional saturation $\bar{Y}_x$ is given by:

$$\bar{Y}_x = \frac{N_x}{4} = \frac{K_{AB}^2(K_{x_B}K_{t_{AB}})[X]+(K_{AB}^4+2K_{AB}^2K_{BB})(K_{x_B}K_{t_{AB}})^2[X]^2}{1+4K_{AB}^2(K_{x_B}K_{t_{AB}})[X]+2(K_{AB}^4+2K_{AB}^2K_{BB})(K_{x_B}K_{t_{AB}})^2[X]^2}$$

$$\frac{+3K_{AB}^2K_{BB}^2(K_{x_B}K_{t_{AB}})^3[X]^3+K_{BB}^4(K_{x_B}K_{t_{AB}})^4[X]^4}{+4K_{AB}^2K_{BB}^2(K_{x_B}K_{t_{AB}})^3[X]^3+K_{BB}^4(K_{x_B}K_{t_{AB}})^4[X]^4} \tag{98}$$

The number of parameters as written in Eqs. (97) and (98) is
four: K_{AB}, K_{BB}, K_{x_B}, and $K_{t_{AB}}$. In fact. Eqs. (97) and (98) contain only two independent parameters (Pauling, 1935) which can
be defined as follows:

$$a = \frac{K_{AB}^2}{K_{BB}} \tag{99}$$

and

$$b = K_{BB}K_{x_B}K_{t_{AB}} \tag{100}$$

Using these definitions, Eq. (98) obtains the form:

$$\bar{Y}_x = \frac{ab[X]+(2ab^2+a^2b^2)[X]^2+3ab^3[X]^3+b^4[X]^4}{1+4ab[X]+2(2ab^2+a^2b^2)[X]^2+4ab^3[X]^3+b^4[X]^4} \tag{101}$$

Equation (101) reveals the important parameter $\dfrac{K_{AB}^2}{K_{BB}}$, which ap-

pears as the *sole* parameter in the dimer case. The parameter $K_{BB}K_{x_B}$ is directly related to the midpoint of the saturation curve. Introducing $\bar{Y}_x = \frac{1}{2}$ into Eq. (101) one can solve the value for $[X_{0.5}]$ and obtain:

$$[X_{0.5}] = \frac{1}{b} \tag{102}$$

It is thus apparent that the cooperativity at 50% saturation ($\bar{Y}_x = \frac{1}{2}$) depends only on *one* parameter namely: K_{AB}^2/K_{BB}. Solving the derivative:

$$n_H = \left(\frac{\partial \log \dfrac{\bar{Y}_x}{1-\bar{Y}_x}}{\partial \log X}\right)_{[X_{0.5}]} \tag{103}$$

namely the Hill coefficient at 50% saturation for the square KNF case one obtains the expression:

$$n_H = \frac{4\left(\dfrac{K_{AB}^2}{K_{BB}} + 1\right)}{\left(\dfrac{K_{AB}^2}{K_{BB}}\right)^2 + 6\dfrac{K_{AB}^2}{K_{BB}} + 1} \tag{104}$$

or

$$n_H = \frac{4(a + 1)}{a^2 + 6a + 1} \tag{105}$$

Thus, knowing the Hill coefficient at the midpoint ($\bar{Y}_x = \frac{1}{2}$) one can directly obtain the value of (K_{AB}^2/K_{BB}).

56

2. *The "Tetrahedral" Case.* In Figure 11 the binding of ligand to a tetrahedral molecule is shown. In this case the ligand-bound species are given by the following expressions:

$$[EX] = [A_3BX] = 4K_{AB}^3 (K_{x_B} K_{t_{AB}})[A_4][X] \tag{106}$$

$$[EX_2] = [A_2B_2X_2] = 6K_{AB}^4 K_{BB}(K_{x_B} K_{t_{AB}})^2 [A_4][X]^2 \tag{107}$$

$$[EX_3] = [AB_3X_3] = 4K_{AB}^3 K_{BB}^3 (K_{x_B} K_{t_{AB}})^3 [A_4][X]^3 \tag{108}$$

$$[EX_4] = [B_4X_4] = K_{BB}^6 (K_{x_B} K_{t_{AB}})^4 [A_4][X]^4 \tag{109}$$

The expressions for ψ_i and the intrinsic association constants K_i' are shown in Table 1.

The binding equation for the tetrahedral model is given by:

$$N_x = \frac{4K_{AB}^3 (K_{x_B} K_{t_{AB}})[X] + 12K_{AB}^4 K_{BB}(K_{x_B} K_{t_{AB}})^2 [X]^2}{1 + 4K_{AB}^3 (K_{x_B} K_{t_{AB}})[X] + 6K_{AB}^4 K_{BB}(K_{x_B} K_{t_{AB}})^2 [X]^2}$$

$$\frac{+12K_{AB}^3 K_{BB}^3 (K_{x_B} K_{t_{AB}})^3 [X]^3 + 4K_{BB}^6 (K_{x_B} K_{t_{AB}})^4 [X]^4}{+4K_{AB}^3 K_{BB}^3 (K_{x_B} K_{t_{AB}})^3 [X]^3 + K_{BB}^6 (K_{x_B} K_{t_{AB}})^4 [X]^4} \tag{110}$$

and the fractional saturation $\bar{Y}_x$ is given by:

$$\bar{Y}_x = \frac{N_x}{4} = \frac{K_{AB}^3 (K_{x_B} K_{t_{AB}})[X] + 3K_{AB}^4 K_{BB}(K_{x_B} K_{t_{AB}})^2 [X]^2}{1 + 4K_{AB}^3 (K_{x_B} K_{t_{AB}})[X] + 6K_{AB}^4 K_{BB}(K_{x_B} K_{t_{AB}})^2 [X]^2}$$

$$\frac{+3K_{AB}^3 K_{BB}^3 (K_{x_B} K_{t_{AB}})^3 [X]^3 + K_{BB}^6 (K_{x_B} K_{t_{AB}})^4 [X]^4}{+4K_{AB}^3 K_{BB}^3 (K_{x_B} K_{t_{AB}})^3 [X]^3 + K_{BB}^6 (K_{x_B} K_{t_{AB}})^4 [X]^4} \tag{111}$$

Equation (111) contains four different parameters but as in the "square" case, the number of independent parameters is only two. These parameters are:

$$a = \frac{K_{AB}^2}{K_{BB}} \tag{112}$$

and

$$g = K_{BB}^{3/2} K_{x_B} K_{t_{AB}} \tag{113}$$

Rewriting Eq. (111) using Eqs. (112) and (113) one obtains:

$$\bar{Y}_x = \frac{a^{3/2}g[X]+3a^3g^2[X]^2+3a^{3/2}g^3[X]^3+g^4[X]^4}{1+4a^{3/2}g[X]+6a^3g^2[X]^2+4a^{3/2}g^3[X]^3+g^4[X]^4} \tag{114}$$

As in the square case, the main parameter which governs the co-operativity of the system is K_{AB}^2/K_{BB}, since g is directly related to the ligand concentration at 50% saturation:

$$[X_{0.5}] = \frac{1}{g} \tag{115}$$

As in the square case the cooperativity at the midpoint is exclusively dependent on K_{AB}^2/K_{BB}. This is expressed in the formula for the Hill coefficient at 50% saturation:

$$n_H = \frac{4\left(\left(\dfrac{K_{AB}^2}{K_{BB}}\right)^{3/2} + 1\right)}{3\left(\dfrac{K_{AB}^2}{K_{BB}}\right)^2 + 4\left(\dfrac{K_{AB}^2}{K_{BB}}\right)^{3/2} + 1} \tag{116}$$

or

$$n_H = \frac{4[a^{3/2} + 1]}{3a^2 + 4a^{3/2} + 1} \tag{117}$$

Thus by measuring the Hill coefficient at the midpoint, the value of K_{AB}^2/K_{BB} can be computed directly.

Both the "square" and the "tetrahedral" models predict a ligand-binding curve which is symmetric about the midpoint in the $\bar{Y}_x$ vs. log[X] plot. Furthermore, most published binding curves when examined demonstrate at first approximation symmetry about the midpoint in the $\bar{Y}_x$ vs. log[X] plot. Extremely accurate binding data is needed in order to establish whether a ligand-binding curve is exactly symmetric about the midpoint in the $\bar{Y}_x$ vs. log[X] plot.

It was already pointed out by Koshland et al. (1966) that the square model and the tetrahedral model can both fit a set of experimental data, since extremely accurate data are needed if one wishes to determine which of the models applies to a real situation.

3. Michaelis-Menten Case. As in the dimer case a Michaelian saturation curve (noncooperative) is obtained when $K_{AB}^2/K_{BB} = 1$. When such a relationship holds, it does not necessarily mean that $K_{AB} = 1$ and $K_{BB} = 1$, which will be a special case. The equality of $K_{AB}^2/K_{BB} = 1.0$ means that a unique relationship between the relative stabilities of the protein conformations predicts a noncooperative binding curve.

5. *The Influence of the Intersubunit Binding Domains on the Nature of Subunit Interactions*

1. General Formulations. In an oligomeric protein composed of identical subunits, a subunit in contact with two neighbors will inevitably have a different set of amino acid residues in contact with neighbor 1 from the set in contact with neighbor 2. This is a necessary consequence of the chirality of protein subunits. These intersubunit domains are designated as p, q and r, etc. In the original mathematical derivation of the sequential model (Koshland et al., 1966), this complexity was avoided by using averaged subunit interaction terms, namely by assuming that every interaction is of the same type. This approach is a simplification and in fact the complete model should take into account

the different types of intersubunit domains within the oligomeric structure. Such an extension of the model was provided by
Cornish-Bowden and Koshland (1970). Since many oligomeric proteins are tetramers composed of identical subunits, we shall
analyze the sequential model in detail only for the case of
tetramers. A similar procedure can be used for higher oligomeric
structures and for oligomers composed of nonidentical promoters.

From Table 1 it can be seen that in the tetramer case when only
one type of K_{AB_2} and one type of K_{BB} term are considered, the
relation $\psi_3^2 = \psi_1 \psi_4$ or $K_1' K_4' = K_2' K_3'$ holds for both the tetrahedral
and the square cases. It is interesting that in the case where
three types of intersubunit domains, (pp, qq, rr) are considered,
in a tetrahedral oligomer possessing 2:2:2 symmetry, the same
relation: $K_1' K_4' = K_2' K_3'$ holds. It is also clear that in the
"square" case where only two types of intersubunit domain are
considered, this relationship also holds. This fact can be easily
verified by looking at the ψ_i values for the square (I_4) and for
the tetrahedral (I_6) cases (Table 2).

Table 2. Values of ψ_1, ψ_2, ψ_3 and ψ_4 for the general square and the general
tetrahedral KNF models

	Square	Tetrahedral
Definitions:	$K_{\overline{AB}} = (K_{ABpp} K_{ABqq})^{1/2}$ $K_{\overline{BB}} = (K_{BBpp} K_{BBqq})^{1/2}$	$K_{\overline{AB}} = (K_{ABpp} K_{ABqq} K_{ABrr})^{1/3}$ $K_{\overline{BB}} = (K_{BBpp} K_{BBqq} K_{BBrr})^{1/3}$
ψ_1	$4K_{\overline{AB}}^2 (K_{x_B} K_{t_{AB}})$	$4K_{\overline{AB}}^3 (K_{x_B} K_{t_{AB}})$
ψ_2	$\left[2K_{\overline{AB}}^2 [K_{BBpp}+K_{BBqq}] + 2K_{\overline{AB}}^4 \right] (K_{x_B} K_{t_{AB}})^2$	$2[K_{ABpp}^2 K_{ABqq}^2 K_{BBrr} + K_{ABpp}^2 K_{ABrr}^2 K_{ABqq}$ $+ K_{ABqq}^2 K_{ABrr}^2 K_{ABpp}] (K_{x_B} K_{t_{AB}})^2$
ψ_3	$4K_{\overline{AB}}^2 K_{\overline{BB}}^2 (K_{x_B} K_{t_{AB}})^3$	$4K_{\overline{AB}}^3 K_{\overline{BB}}^3 (K_{x_B} K_{t_{AB}})^3$
ψ_4	$K_{\overline{BB}}^4 (K_{x_B} K_{t_{AB}})^4$	$K_{\overline{BB}}^6 (K_{x_B} K_{t_{AB}})^4$

The binding equation for a tetrahedral protein possessing the three types of binding domains p, q, and r can be written in terms of ψ_1, ψ_2, ψ_3, and ψ_4, as depicted in Table 2.

Also given in the table are the values of ψ_1, ψ_2, ψ_3, and ψ_4 for the case of a square model possessing only two types of subunit interactions.

As in the case of the simplified tetrahedral protein, the binding curve predicted by the model assuming three types of subunit interactions is symmetric about the midpoint. The midpoint ligand concentration is given by:

$$[X_{0.5}] = (K_{BBpp}K_{BBqq}K_{BBrr})^{-\frac{1}{2}}(K_{x_B}K_{t_{AB}})^{-1} \tag{118}$$

and the Hill coefficient at 50% saturation is given by:

$$n_H = \frac{4(def + 1)}{4def + d^2e^2 + d^2f^2 + e^2f^2 + 1} \tag{119}$$

where:

$$d = (K_{ABpp}^2 K_{BBqq})^{\frac{1}{2}} \tag{120}$$

$$e = (K_{ABqq}^2/K_{BBqq})^{\frac{1}{2}} \tag{121}$$

and

$$f = (K_{ABrr}/K_{BBrr})^{\frac{1}{2}} \tag{122}$$

It can be seen that the Hill coefficient at 50% saturation depends on the quantity $(K_{AB_i}^2/K_{BB_i})^{\frac{1}{2}}$ as in the case of the simplified tetrahedral protein.

The binding of ligand to a tetrameric protein depends on four independent parameters, namely the Adair constants. The symmetry condition applied for the midpoint of the binding curve reduces

the number of independent parameters controlling cooperativity
to two, since:

$$[X_{0.5}] = (\psi_4)^{-\frac{1}{4}}$$

and

$$\psi_3^2 = \psi_1^2 \psi_4 \qquad (123)$$

when $\bar{Y}_X$ vs. $\log[X]$ is symmetric about the midpoint.

These relationships were derived in Chapter 3 Section X [Eqs.
(37) to (40)].

It is sometimes possible to determine experimentally the ratio
between the ligand affinity of the last binding step to the affinity of the first binding step, namely, the ratio K_4'/K_1'. This
can be done by extrapolating the Hill plot at both very low and
very high substrate concentration [see Fig. 4 and Eq. (49)]. The
value K_4'/K_1' is related to the values of K_{AB}^2/K_{BB} by the follow-.
ing expression

$$\frac{K_4'}{K_1'} = \left(\frac{K_{ABpp}^2}{K_{BBpp}} \times \frac{K_{ABqq}^2}{K_{BBqq}} \times \frac{K_{ABrr}^2}{K_{BBrr}} \right)^{-1} \qquad (124)$$

or

$$\frac{K_4'}{K_1'} = def \qquad (125)$$

Thus both the Hill coefficient at 50% saturation [Eq. (119)]
and the ratio K_4'/K_1' are defined in terms of the three $K_{AB_i}^2/K_{BB_i}$
terms. It is therefore clear that ligand-binding experiments
alone cannot determine the value of the three parameters, describing subunit interactions in all three directions of space.

2. Simplified Formulations. The binding scheme described by the
Koshland-Némethy-Filmer model for the tetrahedral case is, in
fact, the most general description of ligand binding to a tetrameric protein. The simple tetrahedral model analyzed originally
by Koshland et al. (1966) assumes that all three subunit contacts are identical, namely that:

$$d = e = f \tag{126}$$

and Eq. (119) reduces to Eq. (117). The simple square model assumes that only two of the three interactions are important and
equal to each other. Namely, in this case the interaction term
$f = 1$ and $d = e$. In this situation Eq. (119) reduces to Eq.
(105).

A further simplification of the tetrahedral model is the assumption that only the interactions across one intersubunit domain play a role in the cooperativity. In this later case:

$$n_H = \frac{4(f + 1)}{4f + 2f^2 + 2} = \frac{2}{f + 1} = \frac{2}{1 + \sqrt{\dfrac{K_{AB}^2}{K_{BB}}}} \tag{127}$$

and the midpoint is given by:

$$[X]_{0.5} = K_{BB}^{-\frac{1}{2}} (K_{x_B} K_{t_{AB}})^{-1} \tag{128}$$

Relations (127) and (128) are identical to the equations describing the dimer [see Sec. III.1 of this Chapter; Eq. (92)].

In fact, the equations describing the dimer apply to all oligomeric proteins in which the cooperative unit is an isologous
dimer, namely, in situations where only one set of the intersubunit domains determines the cooperative behavior of the protein (see Ch. 3).

IV. The Conformational State of the Protein

We have indicated above that the binding of a ligand to an oligomeric protein results in a conformational change. The conformational change associated with ligand binding is not restricted to the liganded subunit, and occurs also in unliganded subunits. This is due to the fact that conformational changes may propagate from the ligand-bound subunit through the intersubunit binding domains. It is therefore of interest to study, in parallel to ligand binding, the conformational changes which occur in the protein. Monod et al. (1965) have defined a function called the state function, $\bar{R}$. This function, according to the original definition, is the fraction of the protein which is in the R conformation. For the sake of simplicity let us examine the qualitative behavior of the state function during the binding of ligand to a protein dimer. Let us examine four cases: (a) the exclusive binding in the MWC model (b) nonexclusive binding in the MWC model (c) the simple sequential KNF model (d) the general sequential KNF model.

1. *Exclusive Binding in the MWC Model*

From Figure 7 it is apparent that the state function $\bar{R}$ will be given by:

$$\bar{R} = \frac{\square\square + \boxed{x}\,\square + \boxed{x}\,\boxed{x}}{\bigcirc\bigcirc + \square\square + \boxed{x}\,\square + \boxed{x}\,\boxed{x}} \tag{129}$$

where a square represents the bound conformation and a circle the unbound conformation. The saturation function will be given by:

$$\bar{Y}_x = \frac{\boxed{x}\,\square + 2\,\boxed{x}\,\boxed{x}}{2\,\bigcirc\bigcirc + 2\,\square\square + 2\,\boxed{x}\,\square + 2\,\boxed{x}\,\boxed{x}} \tag{130}$$

It is immediately clear from Eqs. (129) and (130) that $\bar{R} > \bar{Y}_x$ throughout the titration curve, and if one plots $\bar{R}$ versus $\bar{Y}_x$, one will obtain a concave downward curve (curve B in Fig. 12).

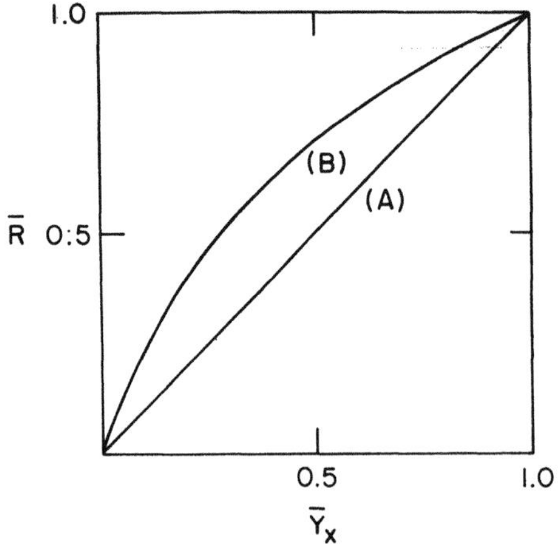

Fig. 12. The dependence of the state function on the saturation function. $\bar{R}$ is the fraction of enzyme subunits in the bound conformation (*square*) as a function of the fractional ligand saturation $\bar{Y}_x$. The MWC model will always generate a curve of the B type. The KNF model yields a curve of the type A in its simplest formulation and a B type curve in its more general formulation. As pointed out in the text, whenever a curve of the A type is encountered it indicates that the conformational change occurring upon ligand binding follows a sequential pattern. However, when a curve of the B type is encountered it is not possible on the basis of this data alone to indicate which of the models fits better the experimental data. In fact, both the MWC and the KNF models can usually generate any particular B type curve

2. The Nonexclusive Binding in the MWC Model

From Figure 7 it is apparent that the state function $\bar{R}$ will be given by:

$$\bar{R} = \frac{\square\square + \square\boxtimes + \boxtimes\boxtimes}{\bigcirc\bigcirc + \bigcirc\otimes + \otimes\otimes + \square\square + \square\boxtimes + \boxtimes\boxtimes} \tag{131}$$

and the saturation function $\bar{Y}$ is given by:

$$\bar{Y}_x = \frac{\bigcirc\otimes + 2\otimes\otimes + \square\boxtimes + 2\boxtimes\boxtimes}{2\bigcirc\bigcirc + 2\bigcirc\otimes + 2\otimes\otimes + 2\boxtimes\boxtimes + 2\boxtimes\square + 2\boxtimes\boxtimes} \tag{132}$$

In this case again $\bar{R} > \bar{Y}_x$ throughout the titration curve.

3. *The Simple Sequential KNF Model*

From Figure 7 it is apparent that the state function $\bar{R}$ is given by:

$$\bar{R} = \frac{\bigcirc\boxed{x} + 2\,\boxed{x}\boxed{x}}{2\bigcirc\bigcirc + 2\,\bigcirc\boxed{x} + 2\,\boxed{x}\boxed{x}} \tag{133}$$

In this case the saturation function $\bar{Y}_x$ is given by:

$$\bar{Y}_x = \frac{\bigcirc\boxed{x} + 2\,\boxed{x}\boxed{x}}{2\bigcirc\bigcirc + 2\,\bigcirc\boxed{x} + 2\,\boxed{x}\boxed{x}} \tag{134}$$

It is clear that in this case $\bar{Y}_x = \bar{R}$ throughout the titration of protein with ligand and thus the plot of $\bar{R}$ versus $\bar{Y}_x$ in this case yields a straight line with a slope of 45° (Fig. 12A).

4. *The General Sequential KNF Model*

In the general case the conformation of the unbound subunit does not necessarily remain a circle as in the simple case (Fig. 9a). Thus one can assume, for example, that the unbound subunit in the monobound species assumes a conformation different from both the unbound subunit and the bound subunit (Fig. 9b). Further binding causes an additional conformational change where the fully saturated dimer is again symmetric. This symmetry in the fully liganded state is not a necessary condition and the possibility that the liganded state is asymmetric in structure is also possible. It is clear that in this case $\bar{R} > \bar{Y}_x$ and the shape of the $\bar{R}$ versus $\bar{Y}_x$ plot in concave upward much in the same way as predicted by the MWC model (Fig. 12B).

From the four cases discussed above, it is apparent that whenever $\bar{R} = \bar{Y}_x$, only the sequential model can account for it. However, whenever $\bar{R} > \bar{Y}_x$, both the MWC and the KNF models are capable of explaining this relationship. The arguments described here for a dimer are equally valid for higher oligomers.

5. Measuring $\bar{R}$

We have discussed the function $\bar{R}$ without specifying the experimental approach to measure the state function. Many experimental techniques are available to the researcher to measure the conformational changes which occur upon ligand binding to a protein. These include, for example, quenching of the protein fluorescence upon ligand binding as in the case of NAD^+ binding to glyceraldehyde-3-phosphate dehydrogenase (Schlessinger and Levitzki, 1974; Henis and Levitzki, 1977). In the latter case it was pointed out that the shape of the $\bar{R}$ function depends upon whether one monitors the quenching of "blue" tryptophans or of "red" tryptophans (Schlessinger and Levitzki, 1974). It has been demonstrated (Schlessinger and Levitzki, 1974; Henis and Levitzki, 1977) that when the fluorescence of tryptophans emitting at higher wavelength is monitored as a function of NAD^+ occupancy, $\bar{R}$ is linearly dependent on $\bar{Y}_x$. If, however, the fluorescence of tryptophans emitting at lower wavelengths is monitored as a function of NAD^+ occupancy, the dependence of $\bar{R}$ on $\bar{Y}_x$ is found to be concave upward (Schlessinger and Levitzki, 1974). It is thus apparent that the "red" tryptophans monitor conformational changes which occur exclusively within the subunit which binds the ligand, and not conformational changes which are also transmitted to neighboring vacant subunits. "Blue" tryptophans on the other hand seem to monitor conformational changes which occur both in the subunit which binds ligand as well as in the vacant neighboring subunits. From this example it is apparent that the shape and behavior of the $\bar{R}$ function depends on the method of measurement. Many other techniques are available to measure the state function $\bar{R}$. For example the change in sedimentation coefficient as a function of ligand binding has been used in the case of an aspartate transcarbamylase (Kirschner and Schachmann, 1973) and references therein). In the case of hemoglobin, the spectroscopic properties of the iron within the heme moiety can be monitored (Perutz, 1972, and references therein) as a function of O_2 binding or of CO binding. For the latter case both optical properties and magnetic properties of the iron can be monitored. Other spectroscopic methods used are changes in the difference spectrum in the absorption of specific chromophores attached to the enzyme as a function of ligand binding. Thus for example the change in the spectrum of a mitrotyrosine within the aspartate transcarbamylase was used to monitor conformational changes which occur upon ligand binding (Kirschner and Schachmann, 1973). Chemical and biochemical tools were also used to monitor protein conformational changes which occur upon ligand binding. For example the susceptibility of a glyceraldehyde-3-phosphate dehydrogenase to proteolytic digestion by trypsin was monitored as a function of NAD^+ occupancy (Fenselau, 1970). Similar experiments were reported on numerous other enzymes. No attempt will be made here to review all the methods which are used to explore the nature of the state function $\bar{R}$. The point is only made that the nature of $\bar{R}$ depends on the *method* used since every method explores a different area of the protein molecule which responds differently to ligand binding. Therefore generalizations on the relationship between the $\bar{R}$ function and the $\bar{Y}_x$ function are at best approximate. In order to explore the nature of the conformational changes occurring in a protein upon ligand binding one must use *a number of* chemical and physicochemical methods. Such a comprehensive approach has been used with great success by Perutz (1972) in his studies on hemoglobin. The current amount of knowledge on the conformational effects of ligand binding in a few other enzymes such as aspartate transcarbamylase and glyceraldehyde-3-phosphate dehydrogenase is also fairly substantial.

V. Comparison Between the KNF Model and the MWC Model

Both the MWC and the KNF models deal with oligomeric structures which are symmetric in design. The MWC model assumes that the symmetry in the oligomeric structure is conserved during the action of the protein and once a conformational change occurs upon ligand binding all the subunits change their conformation in a concerted fashion. Thus unliganded subunits always possess a conformation identical to the liganded subunits. The KNF model, on the other hand, allows for hybrid conformational states where the liganded subunit can assume a conformation which is differ-

ent from the unliganded one. This difference accounts for the
fact that the KNF model allows for negative cooperativity, whereas the MWC model is incapable of doing so.

The KNF model, in contrast to the MWC model, defines and examines
in detail the parameters characterizing the intersubunit interactions. It accounts for the fact that subunit interactions differ in the different directions of space, whereas the MWC model
fails to take this structural element into account.

In general, most positively cooperative binding curves can be
fitted by both the KNF and the MWC models, and thus on the basis
of binding data alone it is usually difficult to decide which
of the models should be applied to describe the data. Other experimental approaches, besides binding experiments, should be
used in such cases. These approaches are mainly centered around
the elucidation of the conformational state of the subunits
during ligation, and mainly involve spectroscopic measurements.
Only in the case of negative cooperativity and mixed positive-
negative cooperativity in ligand binding one can explicitly state
that the KNF model should be used, since the MWC model does not
allow for negative cooperativity. It is worth noting that the
model-independent approach, using the Adair equation (Ch. 3)
allows for negative cooperativity. Thus it seems that the general sequential model of Koshland-Némethy and Filmer represents
a more comprehensive and realistic approach. Still, in many
cases the model of Wyman, Monod and Changeux is a very useful
one, because of its simplicity and elegance. It should be noted
that whenever very high cooperativities are encountered, an almost concerted conformational transition occurs upon ligand
binding. In this case the contribution of intermediate conformational states to the overall binding scheme is insignificant.
Thus the KNF model in these extreme circumstances extrapolates
to the MWC model.

One can eliminate the restriction of the MWC model which calls
for the conservation of symmetry during ligand binding. Under
these conditions the same oligomeric structure can possess subunits in the R-state and the T-state simultaneously. The use of

such a hybrid model was proposed by Kirschner et al. (1966). In
fact this hybrid model is no different from the simplified sequential model published originally by Koshland et al. (1966).
The difference between the two is in the definition of parameters and is therefore semantic in nature.

It is my feeling that the use of the general sequential model
or the modified MWC scheme, allowing for asymmetry in the molecule, is the most fruitful approach to the study of the behavior
or regulatory macromolecules.

The case of hemoglobin may be a good example where a hybrid
model should be used. Let us therefore examine the current pic-
ture of the allosteric transitions in hemoglobin which occur
upon oxygenation. Perutz, through his X-ray crystallographic
studies and spectral measurements, proposed a model which can
be termed semi-concerted, and in fact incorporates many elements
of the MWC concerted model, and some elements from the KNF sequential model. One can write the sequence of hemoglobin oxygenation and the corresponding conformational changes as folows:

$$[\alpha^t\alpha^t\beta^t\beta^t]^T \xrightarrow{O_2} [\alpha^r_{O_2}\alpha^t\beta^t\beta^t]^T \xrightarrow{O_2} [\alpha^r_{O_2}\alpha^r_{O_2}\beta^t\beta^t]^T$$

$$L \Updownarrow$$

$$[\alpha^r_{O_2}\alpha^r_{O_2}\beta^t\beta^t]^R$$

$$\Updownarrow$$

$$[\alpha^r_{O_2}\alpha^r_{O_2}\beta^r\beta^r]^R$$

$$O_2 \Updownarrow$$

$$[\alpha^r_{O_2}\alpha^r_{O_2}\beta^r_{O_2}\beta^r]^R$$

$$O_2 \Updownarrow$$

$$[\alpha^r_{O_2}\alpha^r_{O_2}\beta^r_{O_2}\beta^r_{O_2}]^R$$

Scheme 1. Binding of O_2 to hemoglobin

Small t denotes the tertiary structure of the subunit characteristic for the low affinity deoxy state. Small r denotes the

tertiary structure of the subunit characteristic for the high affinity state of the subunit. T denotes the quaternary conformation of the hemoglobin tetramer characteristic for the deoxy state. R denotes the quaternary conformation characteristic for the oxy state of the hemoglobin tetramer. For simplicity $\alpha_{O_2}^t$ and $\beta_{O_2}^t$ species were omitted, although a finite affinity for oxygen to the "t" conformation of the subunits can be assigned.

From the scheme it is apparent that a first oxygen binds to the α subunit, inducing a tertiary conformational change in the tertiary structure to the r state. The quaternary conformation of the monoliganded and the di-liganded species is still that of deoxyhemoglobin, namely T. However, in the bi-liganded state a concerted transition occurs such that the quaternary conformation of the di-liganded tetramer can assume the R conformation. This equilibrium between the R and T conformation is characterized by an allosteric constant L. The quaternary R conformation of the hemoglobin tetramer can now facilitate the conversion of the low affinity state of the β subunit to the high affinity state r, thus facilitating further oxygenation. The sequence of events as described in the above scheme allows hybrid conformational states not allowed by the original classical MWC model. Although the transition from the T-state to the R-state is concerted, the overall sequence of events is sequential in nature. The case of hemoglobin is of great interest since it represents a case where a very detailed knowledge on the nature of the conformational transitions which accompany the process of oxygenation has been accumulated. Numerous spectroscopic techniques and kinetic measurements have supplemented the extremely elegant X-ray crystallographic studies, and allowed the formulation of the events which occur in hemoglobin upon oxygenation. It is possible that more details will emerge in the future on the nature of the conformational transitions, but the main finding is that one has to consider a number of conformational states. Furthermore, it is clear that the conformational changes which are coupled to oxygen binding occur in a sequential fashion with one concerted transition.

Looking at the scheme of oxygen binding to hemoglobin, it is quite clear that a quantitative description of oxygen binding is most conveniently described by the KNF sequential model. One can, however, define a set of parameters using the Monod-Wyman-Changeux nomenclature. The scheme as written above uses the notations of the MWC model. Since symmetry is not conserved during the process of oxygenation, it becomes a matter of semantics which of the set of notations is used to describe the process of oxygen binding to hemoglobin.

Chapter 6
Special Types of Cooperative Systems

I. Cooperativity Resulting from Ligand-Coupled Protein Association or Dissociation

As we have learned, cooperativity in multisubunit proteins is due to changes in the strength of subunit interactions coupled to ligand binding. The modulation of subunit interactions by ligand binding brings about positive cooperativity, negative cooperativity and mixed-type cooperativity. Some proteins associate and others dissociate upon ligand binding (Duncan et al., 1972; Levitzki and Koshland, 1972, and references therein). In such cases, ligand binding is always positively cooperative, as was already pointed out by Klotz et al. (1970) and Frieden (1971). It is quite clear that fairly complicated ligand-binding curves can be obtained whenever association-dissociation phenomena are coupled to ligand binding (Frieden, 1971; Nichol et al., 1967; Levitzki and Schlessinger, 1974; Dahlquist, 1977). Since not too many cooperative systems are known (about a dozen altogether) where association-dissociation phenomena are involved, we shall restrict our discussion to simple cases.

1. Ligand–Coupled Monomer–Dimer Equilibrium

To gain some insight into ligand-dependent protein association or dissociation, we shall examine the behavior of a dimeric structure E_2 which obeys the following scheme:

$$
\begin{array}{ccc}
2E & \overset{K_D}{\rightleftharpoons} & E_2 \\
K_1 \updownarrow & & \updownarrow K_2 \\
EX & & E_2X \\
& & \updownarrow K_2 \\
& & E_2X_2
\end{array}
$$

Scheme 2. Dimer-monomer equilibrium coupled to ligand binding

$$K_D = \frac{[E_2]}{[E]^2} \tag{135}$$

The monomer form of the enzyme binds the ligand X with an association constant K_1 and the dimeric enzyme has a ligand affinity constant K_2. For simplicity we shall assume that the sites on the dimer are identical and noninteracting. Such a scheme can generate cooperativity in ligand binding by two ways:

a) If K_D is large, so that most of the protein exists as a dimer but the monomer binds the ligand tighter than the dimer ($K_1 > K_2$). Under these conditions ligand binding will be positively cooperative and ligand binding drives the dissociation of the dimer.

b) If K_D is small, and $K_2 > K_1$, the unliganded enzyme, exists largely as a monomer, ligand binding is again positively cooperative and promotes dimer formation.

Let us examine the cooperativity of such a system. The saturation function $\bar{Y}_x$ will be given by:

$$\bar{Y}_x = \frac{[EX] + [E_2X] + 2[E_2X_2]}{[E] + [EX] + 2[E_2] + 2[E_2X] + 2[E_2X_2]} \tag{136}$$

From the above scheme it is apparent that:

$$\bar{Y}_x = \frac{K_1[X] + 2K_DK_2[E][X] + 2K_DK_2^2[E][X]^2}{1 + K_1[X] + 2K_D[E](1 + K_2[X])^2} \tag{137}$$

and

$$[E_o] = [E] + [EX] + 2[E_2] + 2[E_2X] + 2[E_2X_2] \tag{138}$$

$$[E_o] = (1 + K_1[X])[E] + 2K_D[E]^2(1 + K_2[X])^2 \tag{139}$$

Using the definition of the Hill slope, we obtain the formula:

$$n = \frac{d\ln \dfrac{\bar{Y}_x}{1 - \bar{Y}_x}}{d\ln[X]} = [X]d\ln \frac{\dfrac{\bar{Y}_x}{1 - \bar{Y}_x}}{dx}$$

$$= \frac{[X]d\ln \dfrac{K_1[X]+2K_D K_2[E][X]+2K_D K_2^2[E][X]^2}{1+2K_D[E]+2K_D K_2[E][X]^2}}{dx} \tag{140}$$

The expression for n in this case differs from those obtained
for systems which show no association or dissociation in that
the Hill coefficient characterizing the association-dissociation
system depends on the total protein concentration. Thus, both
the position and shape of the saturation curve, as well as of
the Hill plot, depends on the total protein concentration $[E_o]$
used in the experiment. Therefore whenever the $\bar{Y}_x$ function and
the Hill coefficient are found to depend on total protein concentration, it is a good diagnostic indication that ligand binding is coupled to an association or dissociation of the binding protein. The analytical solutions describing the dependence
of $\bar{Y}_x$, as well as of n on [X], are rather complex (Levitzki and
Schlessinger, 1974; Dahlquist, in press), and the best means to
learn about the features of n or $\bar{Y}_x$ is by numerical solutions
using the computer, as was done by Levitzki and Schlessinger
(1974) for the general dimer case (where cooperativity is also
allowed within the dimer species). A scheme which assumes no
cooperativity within the dimer allows in fact only for positive
cooperativity. In the more general model where the two binding
steps describing the formation of E_2X and E_2X_2 respectively are
not characterized necessarily by the same affinity constant,
negative cooperativity as well as mixed positive-negative cooperativity, can be encountered (Levitzki and Schlessinger, 1974).
It is, however, clear that if one wishes to explore the effect
of association-dissociation only, on the cooperativity observed
in ligand binding, it is more beneficial to examine the simplified scheme as represented in Scheme 2. A few generalizations
on the behavior of such a system can be made: (a) the midpoint

concentration $[X_{0.5}]$ depends strongly on the total enzyme concentration, (b) the maximal Hill coefficient (the slope of the Hill plot) occurs away from the midpoint: in associating systems the Hill slope will tend to unity at high fractional saturation, and tend toward two at low fractional saturation. In dissociating systems the reverse situation occurs, (c) the Hill plot displays distinct nonlinearity, and is usually concave downward. To gain more insight into the behavior of ligand-dependent association-dissociation, let us examine two extreme cases.

Case I. Ligand binds exclusively to the dimeric from, monomers exist predominantly in the absence of ligand ($K_D[E_o] \ll 1$; $K_1 = 0$). In this case $K_1 = 0$, and K_D is rather small; namely, the unbound species is almost exclusively a monomer, and the bound species almost exclusively a dimer. This case, therefore, represents ligand-dependent association. Under these conditions the ligand saturation function is given by:

$$\bar{Y}_x = \frac{2K_D K_2 [E][X](1 + K_2[X])}{1 + 2K_D[E](1 + K_2[X])^2} \tag{141}$$

and

$$[E_o] = [E] + 2K_D[E]^2(1 + K_2[X])^2 \tag{142}$$

If very little enzyme exists in the unbound state near the midpoint, one can obtain from Eqs. (141) and (142) that:

$$[X_{0.5}] = \frac{1}{K_2}\left(\frac{1}{K_D[E_o]}\right)^{\frac{1}{2}} \tag{143}$$

when $K_2[X] \gg 1$. Namely, the midpoint ligand concentration of the saturation curve is inversely dependent on the square root of the total enzyme concentration. Under these conditions ($K_2[X] \gg 1$) the Hill slope is given by:

$$n = 2 - \frac{2\bar{Y}_x}{1 + \bar{Y}_x} \tag{144}$$

Equation (144) was derived by Dahlquist (in press), and its derivation is given in the Appendix. From Eq. (144) it is readily seen that:

$$\lim_{\bar{Y}_x \to 0} n = 2 \qquad (145)$$

and

$$\lim_{\bar{Y}_x \to 1} n = 1 \qquad (146)$$

The Hill plot therefore is concave downward and its maximal slope occurs at vanishingly low ligand occupancy. At 50% saturation $n_H = 1.33 [\bar{Y} = \frac{1}{2}$ in Eq. (144)].

Case II. Ligand binding only to monomeric form ($K_D[E_o] \gg 1$; $K_2 = 0$) In this case the unbound species is almost exclusively a dimer, and the bound species almost exclusively the protein monomer. This case, therefore, represents ligand-dependent dissociation.

We therefore consider a case where in the absence of ligand the dimeric form is favored, but only the monomer binds ligand. Under these conditions the saturation function is given by:

$$\bar{Y}_x = \frac{K_1[X]}{K_1[X] + 2K_D[E]} \qquad (147)$$

and

$$[E_o] = K_1[X][E] + 2K_D[E]^2 \qquad (148)$$

It can be shown that the expression for the Hill slope is:

$$n = 1 + \frac{\bar{Y}_x}{2 - \bar{Y}_x} \qquad (149)$$

Equation (149) was originally derived by Dahlquist (in press), and its derivation is given in the Appendix. In this case, at low degrees of saturation (low $\bar{Y}_x$ values) the Hill coefficient approaches unity, whereas at high degrees of ligand saturation the value of the Hill coefficient approaches two:

$$\lim_{\bar{Y}_x \to 0} n = 1.0 \tag{150}$$

$$\lim_{\bar{Y}_x \to 1} n = 2.0 \tag{151}$$

It is also clear from Eq. (149) that n_H (n at $\bar{Y}_x = \frac{1}{2}$) cannot exceed 1.33.

2. The General Case

The behavior as discussed in the above two limiting cases is characteristic also to associating-dissociating systems composed of many subunits. Let us consider a few major characteristics of such a system.

1. Association. Let us consider a protein composed of p subunits in the absence of ligand, and which associates to a structure composed of ℓ subunits upon complete ligation where each subunit binds a ligand molecule. The maximal Hill coefficient of the midpoint is given by:

$$n_H^{max} = \frac{2p\ell}{p + 1} \tag{152}$$

This relationship is derived under conditions where the dissociated species predominates in the absence of ligand, and the associated species predominates in the presence of ligand. The midpoint ligand concentration is given by:

$$[X_{0.5}] = A \left(\frac{1}{[E_o]} \right)^{\frac{p-1}{p \cdot \ell}} \tag{153}$$

where A is a constant (Dahlquist, in press).

It can be seen that $n_H^{max} = 4/3$ (1.33) in the case of a monomer ($p = 1$) associating into a dimer ($p\ell = 2$). Thus the maximal Hill coefficient at the midpoint is much smaller than 2.0, the limiting Hill coefficient at very low ligand concentration [Eq. (145)]. In the case of a dimer ($p = 2$) going to a tetramer ($p\ell = 4$) upon association $n_H^{max} = 8/3 = 2.67$. Equations (152) and (153) were originally derived by Dahlquist (in press) and their derivation is given in the Appendix.

2. Dissociation. The Hill coefficient at the midpoint, n_H, for a protein oligomer undergoing ligand-dependent dissociation to liganded monomers, is given by:

$$n_H = 1 + \frac{m - 1}{m + 1} \tag{154}$$

where m is the number of monomer units in the unliganded protein species. From Eq. (154) it is apparent that n_H can never exceed 2.0 at the midpoint for this case. However, the Hill slope will tend to attain values larger than 2.0 at high ligand occupancy and can reach the total number of subunits when $\bar{Y}_x \to 1.0$.

<u>Comments:</u>

The occurrence of positive cooperativity in protein undergoing association-dissociation is rather rare. A list of enzymes displaying cooperativity due to ligand coupled association-dissociation is given in a number of recent publications (Levitzki and Koshland, 1972; Levitzki and Schlessinger, 1974; Dahlquist, in press). The total number of enzymes involved is rather small and is in the neighborhood of a dozen enzymes. Ligand-coupled association-dissociation can be an efficient device for regulatory control. It is interesting that in very cooperative proteins, where the Hill coefficient measured approaches the total number of binding sites, it is found that the protein represents an aggregating system in which the aggregation process is tightly coupled to ligand binding. It was suggested (Levitzki and Schlessinger, 1974) that whenever high degrees of cooperativity had to be attained, strong subunit interactions had to evolve in the protein. The largest change in subunit interactions expected upon ligand binding would indeed be when the subunits are physically separated when unliganded and tightly associated when liganded. Similarly, high cooperativity can be generated when the subunits are tightly associated when unliganded, and completely dissociated when liganded. The two extreme cases generate the highest cooperativity at very low and very high ligand occupancies respectively, as is demonstrated for the two cases analyzed for the dimer.

II. Negative Cooperativity

General Comments. Negative cooperativity (anticooperativity) refers to the phenomenon in multisubunit proteins, in which the ligand-binding affinity decreases as a function of ligand saturation (Levitzki and Koshland, 1969). In the simplest case, the affinity of the protein towards the first ligand molecule is higher than the affinity towards the second, the affinity towards the second ligand molecule is higher than the third, etc. This contrasts markedly with the well-known phenomenon of positive cooperativity, which was initially discovered in the binding of oxygen to the hemoglobin molecule, and found later in many regulatory enzymes. In positive cooperativity, the affinity of the protein toward the ligand increases as a function of ligand saturation, thus leading to a sigmoidal saturation curve. Until recently, this was the only type of cooperativity recognized, and thus cooperativity was synonymous with positive cooperativity.

However, it now appears that negative cooperativity is not only widely distributed in enzymes, but may play an important role in membrane receptors (Levitzki and Koshland, 1976, and references therein). Hence it will be useful to employ the term cooperativity (or homotropic effects) to include all types of subunit interactions and to use positive and negative or mixed to indicate the sequence of affinities during binding of successive ligand molecules.

After the prediction (Koshland et al., 1966) of negative co-operativity came the evidence for the first system to possess this phenomenon, namely the binding of NAD^+ to rabbit muscle glyceraldehyde-3-phosphate dehydrogenase (Conway and Koshland, 1968). Shortly after this finding, many other proteins (Levitzki and Koshland, 1969) were found to possess negative cooperativity. The development of simple diagnostic tests had made it possible to identify numerous proteins which possess this property (Levitzki and Koshland, 1969, 1976).

Recently it has become apparent that negative cooperativity can occur also in a most extreme form, namely in the form of half-of-the-sites reactivity (Levitzki and Koshland, 1976, and references therein). Thus, it is found that a number of multi-subunit enzymes bind ligand to only half of their potential sites. In this case the binding at initial half-of-the-sites is so preferred that binding of the remaining ligands occurs only slightly, since the binding affinity is almost undetectable when ordinary substrate and enzyme concentrations are used.

In Figure 13 four different methods for plotting ligand binding to noncooperative, positively cooperative and negatively co-operate proteins are demonstrated.

On the $\bar{Y}_x$ vs.[X] plot, positive cooperativity is characterized by possessing a sigmoidal shape. The noncooperative, Michaelian binding curve, is a rectangular hyperbola, whereas the negatively cooperative curve appears as a slightly flattened hyperbola. The similarity in appearance of the two latter curves is probably the primary reason that negative cooperativity escaped detection for so long.

On the double reciprocal plot (Fig. 13) noncooperative binding yields a straight line, positively cooperative binding - a concave upward curve and negatively cooperative binding - a concave downward curve.

On the Scatchard plot the noncooperative binding yields a straight line, positively cooperative binding, a concave down-

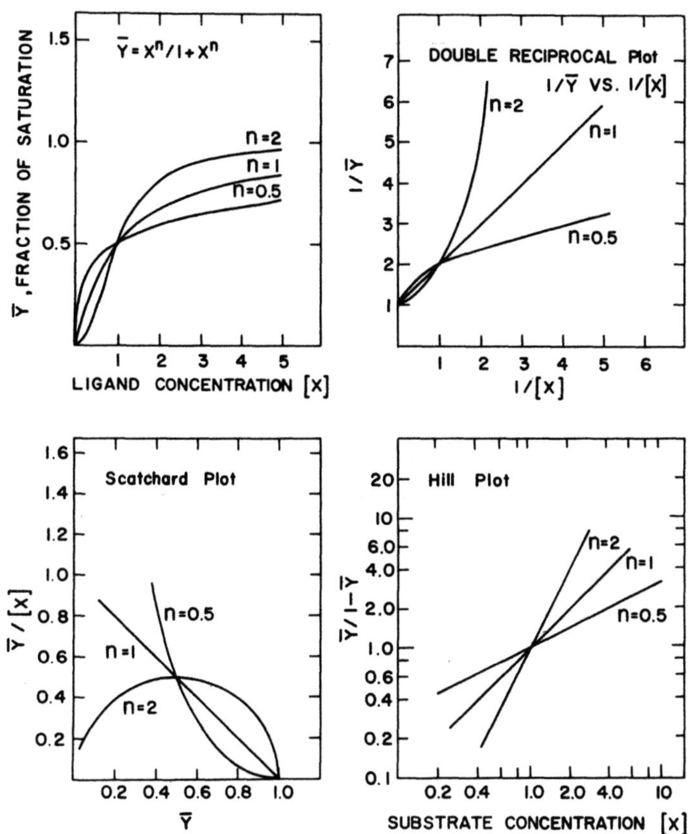

Fig. 13. Noncooperative, positively cooperative and negatively cooperative ligand binding. Four different ways to plot binding data are presented. The specific example used is a protein possessing either noncooperative binding (n_H = 1.0), positively cooperative binding (n_H = 2) and negatively cooperative binding (n_H = 0.5)

ward curve and negatively cooperative binding, a concave upward curve.

On the Hill plot, noncooperative binding yields a Hill coefficient of 1.0, but positive cooperativity a Hill coefficient higher than one, and negative cooperativity a Hill coefficient less than 1.0.

The use of these diagnostic plots has led to the identification of many negatively cooperative systems (Levitzki and Koshland, 1969, 1976).

The evidence from a diagnostic plot does not by itself prove the existence of negative cooperativity. In the first place, it is necessary to establish that the protein sample is pure and that the molecule is composed of identical subunits. Obviously the existence of isozymes or heterogeneity due to protein denaturation, or heterogeneity due to covalent modification of the protein will produce the *appearance* of negative cooperativity in terms of ligand binding curves. Therefore one of the necessary requirements is to establish that a single protein species of identical subunits is being examined.

A second limitation on the use of diagnostic plots of this sort arises from the use of kinetic data. In many cases a pure enzyme is not available and only kinetic measurements can be performed. Under these circumstances many authors have assumed that the extent of ligand saturation $\bar{Y}_x$ can be measured by the ratio $\frac{v}{V_{max}}$, where v is the velocity measured at a certain substrate level and V_{max} is the maximal velocity attainable. Provided that $\frac{v}{V_{max}}$ is a measure for $\bar{Y}_x$, such treatments of the data are acceptable. In the few cases in which $\bar{Y}_x$ was obtained from both direct binding experiments and from kinetic experiments it was established that $\frac{v}{V_{max}}$ is proportional to $\bar{Y}_x$ (Levitzki and Koshland, 1976). However, it is certainly possible to have $\frac{v}{V_{max}}$ being quite different from $\bar{Y}_x$, for example in cases where substrate inhibition occurs. Interestingly, in all of the enzymes studied so far (Levitzki and Koshland, 1969, 1976) negative cooperativity of kinetics has been correlated with negative cooperativity of binding. Hence as a working hypothesis, kinetic experiments are useful as a diagnostic tool but they must always be followed by binding experiments to be sure of the cooperativity pattern.

It is apparent from the analysis of a number of negatively cooperative enzymes (Levitzki and Koshland, 1976) that negative cooperativity is a widespread phenomenon. Furthermore, it is apparent that sometimes the same protein oligomer can exhibit positive cooperativity towards some ligands, and negative cooperativity towards other (Levitzki and Koshland, 1976). These findings are readily explained by the sequential KNF model for subunit interactions. Firstly, the nature of the ligand has a key role in the nature of the conformational change induced, and therefore the type of subunit interactions. Secondly, the nature of the assembly of subunits is also decisive in the way by which the ligand-induced distortions are transmitted. Thus for example, a protein tetramer possessing 2:2:2 molecular symmetry has *three different* intersubunit domains, pp, qq, and rr. Each of the intersubunit domains may function differently, depending on the properties of the domain and on the nature of the bound ligand. Moreover, these relationships can be altered by pH, temperature, or covalent modification. Thus, the extent of conformational change transmitted through the domain and its nature depend on both the domain and the ligand inducing that change. The sequential model can easily deal with a large number of types of intersubunit domains, since the detailed role of intersubunit interaction is taken into account in the basic postulates of the model. Thus the ligand-induced model for subunit interactions (the sequential model) seems to be the simplest molecular model capable of explaining *all* types of cooperativities.

Half-of-the-sites reactivity is also a widespread phenomenon (Levitzki and Koshland, 1976, and references therein). The existence of this phenomenon indicates that the subunit interactions are stronger in one of the isologous intersubunit domains than the other isologous domains in the oligomeric assembly. Thus the binding of a ligand to one subunit will diminish the affinity and/or reactivity of the neighboring subunit across one domain. Other subunits lying in other directions in space within the protein assembly can either be not affected at all or slightly affected. In the former case, the loss of activity as a function of the degree of subsitution of the active sites would be linear. In the case where the interactions across the two other domains are significant, the reactivity of the second site is diminished and the monobound species will react slower with the second molecule of reagent and will exhibit less than 50% activity of the monosubstituted yeast glyceraldehyde-3-phosphate dehydrogenase (Levitzki and Koshland, 1976, and references therein).

Half-of-the-sites reactivity can result in Michaelian steady-state kinetics if the two classes of subunit turnover in a reciprocating fashion. Thus the property of half-of-the-sites reactivity can be missed if no binding studies or pre-steady-state kinetic experiments are performed.

Genuine negatively cooperative binding and half-of-the-sites reactivity may be accounted for in an oligomeric protein essentially by two mechanisms:

1. *Pre-Existing Asymmetry in the Ligand-Binding Sites*. This heterogeneity is manifested by nonidentical affinities toward the ligand and partial sites reactivity. It has been suggested that oligomeric enzymes exhibiting negative cooperativity and half-of-the-sites reactivities are proteins with limited symmetry (Seydoux et al., 1973). For example, tetramers, instead of possessing 2:2:2 symmetry, may possess only 2-symmetry (only one twofold axis of rotational symmetry) where two classes of sites preexist in the absence of ligand or in the presence of ligand. This situation will bring about negatively cooperative ligand binding with *two* classes of affinity and half-of-the-sites reactivity.

2. *Ligand-Induced Sequential Model (KNF)*. In this model the protein oligomer exists in a unique symmetric all-isologous structure in the absence of ligand. Upon ligand binding, conformational changes in the protein are induced and are mainly transmitted through one set of intersubunit domains. Thus, upon binding, two classes of ligand-binding affinities and site reactivities are generated.

In order to evaluate whether a protein is asymmetric in the absence of added ligand, or whether its asymmetry is induced by ligand binding, a number of approaches can be applied:

a) *X-Ray Crystallography*. The structure of a number of oligomeric enzymes exhibiting negative cooperativity or half-of-the-sites reactivity has been studied intensively. It has been found that alkaline phosphatase (Hanson et al., 1970) and liver alcohol dehydrogenase (Bränden et al., 1973) are symmetric dimers in the absence of ligands in the crystal. In these cases therefore, no evidence for pre-existing asymmetry is found in the crystal structure. Such asymmetry in oligomers in the absence of ligand was observed in yeast hexokinase (Steitz et al., 1973) where the asymmetric unit was found to be the dimer. In the crystal, hexokinase is capable of binding only one glucuse molecule per dimer. It is known, however, that the hexokinase dimer dissociates upon glucose binding when the enzyme is in solution to produce ligand-bound monomers. The behavior of the enzyme in solution is therefore rather different from that exhibited by the crystalline enzyme. It is therefore possible that the crystal structure stabilizes an enzyme conformation which is different from the functional one, found in solution. A similar case is the insulin dimer, where it is found to be asymmetric in the crystal (Adams et al., 1969). However, no evidence is available for whether insulin dimers are symmetric or asymmetric when in solution. Recent studies (Moras et al., 1975) on the crystal structure of lobster holoGPDH reveal two classes of NAD^+-binding sites where in two of the four subunits the Cys 149 SH group is hydrogen-bonded to a histidine residue. No data is available as to whether this asymmetry can be observed in the apoenzyme state, or whether it is induced by the binding of NAD^+. Thus crystallography is a valuable probe but it must be remembered that even when asymmetry is observed in the crystal structure it does not immediately follow that functional asymmetry will be found in solution.

b) *Binding Studies.* The pre-existing asymmetry model predicts *two* classes of binding sites and therefore the ability to fit binding curves with two binding constants.

In a number of systems (Levitzki and Koshland, 1976) it is found that ligand binding must be described by more than two binding constants. This situation can only arise if the affinity of the protein towards the ligand decreases progressively as a function of ligand saturation. An alternative explanation would be the existence of a tetramer with four chemically identical subunits, each possessing a different conformation. Such a hypothesis would predict that the protein oligomer should not possess any symmetry element. This situation is not found in any of the proteins investigated.

c) The Dynamics of Active Site Modification. In a number of enzymes such as yeast glyceraldehyde-3-phosphate dehydrogenase and CTP synthetase (Levitzki and Koshland, 1976, and references therein), it has been shown that in the protein tetramer the reactivity of the active site SH groups towards active site directed reagents strongly depends on the number of sites which have already reacted. Thus in yeast glyceraldehyde-3-phosphate dehydrogenase, the four SH groups react *consecutively* with active site reagents where the rate of reaction decreases progressively. A similar situation was observed in the tetrameric form of CTP synthetase where the reaction of the first molecule of the affinity label 6-diazo-5-oxonorleucine (DON) reacts fast, the second one ten times more slowly and the two remaining sites do not react at all. In this latter case it was also possible to *increase* the modification rate of the second site upon adding the allosteric effector GTP (Levitzki et al., 1971). This latter observation demonstrates that half-of the sites reactivity is a property which can be modulated by other specific ligands. This phenomenon has since been observed in other enzymes exhibiting half-of-the-sites reactivity (Levitzki and Koshland, 1976).

In both CTP synthetase, yeast glyceraldehyde-3-phosphate dehydrogenase, and rabbit muscle glyceraldehyde-3-phosphate dehydrogenase, total enzyme activity towards the substrate is lost upon modifying half of the active sites (two). However, the dynamics of the phenomenon of half-of-the-sites reactivity reveals that the reactivity towards the modifying reagent follows a sequential pattern where the site reactivity towards the alkylating agent decreases progressively in *four* steps.

In a recent systematic study on rabbit muscle glyceraldehyde-3-phosphate-dehydrogenase it was shown that the apoenzyme exhibits all-of-the-sites reactivity towards some ligands and half-of-the-sites reactivity towards others. This behavior is undoubtedly due to the fact that the structure of the reagent is crucial in inducing the conformational changes leading to the half-of-the-sites reactivity (Levitzki, 1973, 1974).

d) Conformational Probes. If an oligomeric enzyme is assumed to *pre-exist* in an asymmetric structure, one would expect that ligand binding to one subunit would not cause any conformational changes in the neighboring vacant subunit. In fact, the saturation of tetrameric rabbit muscle apo glyceraldehyde-3-phosphate dehydrogenase with less than *two* NAD$^+$ molecules (Fenselau, 1970), and the phosphorylation of one of the two available sites on succinyl-CoA synthetase (Moffet et al., 1972) offers *full* protection against proteolytic degradation. These enzymes are found to be rapidly digested by the proteolytic enzyme in the absence of ligand. Thus apparently at least in these cases the pre-existent asymmetry model can be ruled out completely.

3. The Biological Significance of Negative Cooperativity. The question which comes to mind immediately is what is the biological significance of negative cooperativity. Negative cooperativity may play a number of roles both in controlling enzyme activity and a whole metabolic pathway. Glyceraldehyde-3-phosphate dehydrogenase occupies a key position in the energy metabolism of the muscle. In muscles, glyceraldehyde-3-phosphate dehydrogenase is extremely negatively cooperative, and the enzyme occurs at very high concentration in the tissue.

It is probably essential that the flow of the glycolytic pathway remains undisturbed even when fluctuations in NAD$^+$ concentrations occur due to other processes occurring in the tissue. The strong negative cooperativity in NAD$^+$ binding insures the insensitivity of the enzyme activity to fluctuations in NAD$^+$ concentration. The enzyme is extremely efficient in binding NAD$^+$, and will not lose the bound coenzyme from its *tight* sites even if the NAD$^+$ concentration drops substantially. The cell, by producing many enzyme molecules, secures the constant availability of enzyme sites with NAD$^+$ bound, in readiness to react with the substrate glyceraldehyde-3-phosphate produced in the preceding steps of glycolysis.

Thus negative cooperativity provides a means of increasing the affinity towards a ligand. This is achieved by increasing the affinity at one site at the expense of the affinities in other sites on the oligomeric protein.

One of the most important roles of negative cooperativity could be the improvement of the catalytic efficiency of one subunit at the expense of some of the ligand-binding energy to the second subunit. For example, in a dimeric protein substrate will bind in a negatively cooperative fashion, where the difference between the ligand affinities is expressed as a rate acceleration in one of the sites. This rate acceleration is brought about by the use of part of the ligand-binding energy at one site to change the protein conformation at a second site, thus improving its catalytic efficiency. Therefore, if the catalytic efficiency in one subunit is increased 10-fold or even 100-fold at the expense of a total loss of activity of half of the subunits, the overall improvement in catalysis is still enormous. Thus half-of-the-sites reactivity may be in fact a way to improve the rate of catalysis.

A more detailed discussion of this point is given elsewhere (Levitzki and Koshland, 1976).

Appendix

I. Obtaining $[X_{0.5}] = \psi_4^{-\frac{1}{4}}$

In the text the relationship

$$[X_{0.5}] = \psi_4^{-\frac{1}{4}} \tag{1}$$

was obtained for a case where the ligand-binding curve is symmetric about the midpoint. This expression is obtained as follows.

The Adair equation for four sites is given by:

$$\bar{Y}_x = \frac{1}{4} \times \frac{\psi_1[X] + 2\psi_2[X]^2 + 3\psi_3[X]^3 + 4\psi_4[X]^4}{1 + \psi_1[X] + \psi_2[X]^2 + \psi_3[X]^3 + \psi_4[X]^4} \tag{2}$$

At the midpoint $\bar{Y} = \frac{1}{2}$; $[X] = [X_{0.5}]$.

Thus Eq. (2) is reduced to:

$$2\psi_4[X_{0.5}]^4 + \psi_3[X_{0.5}]^3 - \psi_1[X_{0.5}] - 2 = 0 \tag{3}$$

This equation cannot be solved analytically for the general case. However, if the Hill plot displays symmetry about the midpoint, namely when the relationship:

$$\psi_1^2\psi_4 = \psi_3^2 \tag{4}$$

holds an expression for $[X_{0.5}]$ can be obtained. Introducing Eq. (4) into Eq. (3) one obtains:

$$2\psi_4[X_{0.5}]4 + \psi_1\sqrt{\psi_4}[X_{0.5}]^3 - \psi_1[X_{0.5}] - 2 = 0 \tag{5}$$

90

which can be rearranged to:

$$2(\psi_4[x_{0.5}]^4 - 1) + \psi_1[x_{0.5}](\sqrt{\psi_4}[x_{0.5}]^2 - 1) = 0 \qquad (6)$$

and

$$2(\sqrt{\psi_4}[x_{0.5}]^2 + 1)(\sqrt{\psi_4}[x'_{0.5}]^2 - 1) + \psi_1[x_{0.5}](\sqrt{\psi_4}[x'_{0.5}]^2 - 1) = 0$$

$$(7)$$

It follows then:

$$(\sqrt{\psi_4}[x_{0.5}]^2 - 1)\ 2(\sqrt{\psi_4}[x_{0.5}]^2 + 1) + \psi_1[x_{0.5}] = 0 \qquad (8)$$

The conditions for Eq. (8) are that either:

$$\sqrt{\psi_4}[x_{0.5}]^2 - 1 = 0 \qquad (9)$$

namely that:

$$[x_{0.5}] = \psi_4^{-\frac{1}{4}} \qquad (10)$$

or

$$2(\sqrt{\psi_4}[x_{0.5}]^2 + 1) + \psi_1[x_{0.5}] = 0 \qquad (11)$$

which results in:

$$x_{0.5} = \frac{-\psi_1 \pm \sqrt{\psi_1^2 - 16\sqrt{\psi_4}}}{4\sqrt{\psi_4}} \qquad (12)$$

Solution (10) is a realistic one. Solution (12) is unrealistic since $[x_{0.5}]$ must be a positive number, since:

$$\sqrt{\psi_1^2 - 16\sqrt{\psi_4}} < \psi_1 \qquad (13)$$

It follows that $[X_{0.5}]$ is a negative quantity according to, relation (13) is not possible. Therefore the only solution is given in Eq. (10) which is identical with Eq. (38) in the text (p. 25).

II. The Relationship of n_H to L and K_R' in the MWC Model (Exclusive Binding)

Let us define

$$y = K_R'[X] \tag{14}$$

thus

$$\ln y = \ln[X] + \ln K_R' \tag{15}$$

and

$$d\ln y = d\ln[X] \tag{16}$$

Therefore one can write:

$$\frac{L + (1 + y)^N}{L + 1} = 1 + \sum \psi_i [X]^i \tag{17}$$

It is clear therefore that

$$\sum i\psi_i [X]^i = \frac{d}{d\ln[X]}(1 + \sum \psi_i X^i) = \frac{d}{d\ln y}\left(\frac{L + (1 + y)^N}{L + 1}\right) \tag{18}$$

one therefore obtains that:

$$\frac{Ny(1 + y)^{N-1}}{L + 1} = \sum i\psi_i [X]^i \tag{19}$$

Similarly one can derive that

$$\frac{N \cdot y}{L + 1}[(1+y)^{N-2}(N-1) \cdot y + (1+y)^{N-1}] = \sum i^2 \psi_i [X]^i \tag{20}$$

$$\frac{N \cdot y}{L + 1} \left[(1+y)^{N-2} \{y(N-1)+1+y\} \right] = \sum i^2 \psi_i [X]^i \tag{21}$$

or

$$\frac{N \cdot y}{L + 1} \left[(1+y)^{N-2} (Ny+1) \right] = \sum i^2 \psi_i [X]^i \tag{22}$$

The Hill coefficient at 50% saturation is given by Eq. (26) of the text:

$$n_H = \frac{4}{N} \left(\frac{\sum i^2 \psi_i [X_{0.5}]^i}{1+ \sum \psi_i [X_{0.5}]^i} - \frac{N^2}{4} \right) \tag{23}$$

Inserting Eqs. (19) and (22) into Eq. (23) one obtains:

$$n_H = \frac{4}{N} \left(\frac{\frac{N \cdot y}{L+1} [1+y]^{N-2} [Ny+1]}{\frac{L+(1+y)^N}{L+1}} - \frac{N^2}{4} \right) \tag{24}$$

therefore:

$$n_H = \frac{4 \cdot y (1 + y)^{N-2} (N \cdot y + 1)}{L + (1 + y)^N} - N \tag{25}$$

it is clear that at the midpoint:

$$\frac{N}{2} = \frac{Ny (1 + y)^{N-1}}{L + (1 + y)^N} \tag{26}$$

namely:

$$L + (1 + y)^N = 2y(1 + y)^{N-1} \tag{27}$$

inserting Eq. (27) into Eq. (25) one obtains:

$$n_H = \frac{4y(1 + y)^{N-2}(N \cdot y + 1)}{2y(1 + y)^{N-1}} - N =$$

$$= \frac{2(Ny + 1)}{y + 1} - N = \frac{Ny + 2 - N}{y + 1} \tag{28}$$

Thus:

$$y = \frac{N + n_H - 2}{N - n_H} \tag{29}$$

but at the midpoint, according to Eq. (14):

$$y = K_R'[X_{0.5}] \tag{30}$$

The value $K_R'[X_{0.5}]$ can be inserted into Eq. (29) and one obtains:

$$K_R'[X_{0.5}] = \frac{N + n_H - 2}{N - n_H} \tag{31}$$

This equation is identical to Eq. (89) in the text.

From Eq. (30) and Eq. (27) one can write:

$$L + (1 + K_R'[X_{0.5}])^N = 2K_R'[X_{0.5}](1 + K_R'[X_{0.5}])^{N-1} \tag{32}$$

or:

$$L = 2K_R'[X_{0.5}](1 + K_R'[X_{0.5}])^{N-1} - (1 + K_R'[X_{0.5}])^N \tag{33}$$

$$L = (1 + K_R'[X_{0.5}])^{N-1}(2K_R'[X_{0.5}] - K_R'[X_{0.5}] - 1) \tag{34}$$

$$L = (1 + K_R'[X_{0.5}])^{N-1}(K_R'[X_{0.5}] - 1) \tag{35}$$

Equation (35) is identical to Eq. (88) in the text.

III. Protein Association and Dissociation Coupled to Ligand Binding

1. Dimerization Coupled to Ligand Binding

Using the equilibrium contants defined in the text, the material balance equation becomes:

$$[E_o] = [E] + [E][X] + 2[E_2] + 2[E_2X] + 2[E_2X_2] \tag{36}$$

$$[E_o] = [E](1+K_1[X]) + 2[E]^2 K_D (1+K_2[X])^2 \tag{37}$$

where $[E_o]$ is the total enzyme concentration.

Solving for E one obtains:

$$[E] = \frac{-(1+K_1[X]) + \sqrt{(1+K_1[X])^2 + 8[E_o]K_D(1+K_2[X])^2}}{4K_D(1+K_2[X])^2} \tag{38}$$

and in the absence of ligand:

$$[E] = \frac{-1 + \sqrt{1 + 8[E_o]K_D}}{4K_D} \tag{39}$$

If K_D is very small, then $[E] = [E_o]$ and the monomer dominates before ligand is added. If K_D is very large the protein exists in the dimeric state before ligands are added.

The expression for the fractional saturation, $\bar{Y}_x$, becomes:

$$\bar{Y}_x = \frac{[EX] + [E_2X] + 2[E_2X_2]}{[E] + [EX] + 2[E_2] + 2[E_2X] + 2[E_2X_2]}$$

$$= \frac{K_1[X] + 2K_D \cdot K_2 \cdot [E][X](1 + K_2[X])}{1 + K_1[X] + 2K_D \cdot [E] \cdot (1+K_2[X])^2} \tag{40}$$

The ratio $\bar{Y}_x/(1 - \bar{Y}_x)$ is given by:

$$\frac{\bar{Y}_x}{1 - \bar{Y}_x} = \frac{K_1[X] + 2K_D[E](1 + K_2[X])(K_2[X])}{+ 2K_D \cdot [E] \cdot (1 + K_2[X])} \tag{41}$$

The slope of the Hill plot at any position is given by the derivative:

$$\frac{d\ln \dfrac{\bar{Y}_x}{1-\bar{Y}_x}}{d\ln[X]} = \frac{K_1[X]+2K_D[E](K_2[X]+2K_2^2[X]^2)+2K_D(1+K_2[X])(K_2[X])\dfrac{d[E]}{d\ln[X}}{K_1[X]+2K_D[E](1+K_2[X])(K_2[X])}$$

$$\frac{-2K_DK_2[E][X]+2K_D(1+K_2[X])\dfrac{d[E]}{d\ln[X]}}{1+2K_D[E](1+K_2[X])} \tag{42}$$

where $\dfrac{d[E]}{d\ln[X]}$ can be evaluated from Eq. (38):

$$0 = (1+K_1[X])\frac{d[E]}{d\ln[X]}+K_1[X]\cdot[E]+4[E]^2K_D(1+K_2[X])(K_2[X])$$

$$+4[E]K_D(1+K_2[X])^2\frac{d[E]}{d\ln[X]} \tag{43}$$

$$\frac{d[E]}{d\ln[X]}=\frac{-[K_1[X]+4[E]K_D(1+K_2[X])(K_2[X])][E]}{1+K_1[X]+4[E]K_D(1+K_2[X])^2} \tag{44}$$

namely:

$$\frac{d\ln[E]}{d\ln[X]}=\frac{-(K_1[X]+4[E]K_D(1+K_2[X])K_2[X]}{1+K_1[X]+4[E]K_D(1+K_2[X])^2} \tag{45}$$

Case 1. Ligand binding only to dimer. Monomer is the predominant species in the absence of ligand.

The conditions require that $K_1 = 0$, $8[E_o]K_D \ll 1$.

Under these conditions:

$$\bar{Y}_x = \frac{2K_D[E](K_2[X])^2}{1+2K_D[E](K_2[X])^2} \tag{46}$$

and

$$\frac{\bar{Y}_x}{1 - \bar{Y}_x} = 2K_D[E](K_2[X])^2 \tag{47}$$

It follows that:

$$n = \frac{d\ln \dfrac{\bar{Y}_x}{1-\bar{Y}_x}}{d\ln[X]} = 4K_D[E](K_2[X])^2 + 2K_D(K_2[X])^2 \frac{d[E]}{d\ln[X]} \tag{48}$$

or:

$$n = 4K_D[E](K_2[X])^2 + 2K_D[E](K_2[X])^2 \frac{d\ln[E]}{d\ln[X]} \tag{49}$$

But when $K_2[X] \gg 1$, Eq. (48) obtains the form:

$$\frac{d\ln[E]}{d\ln[X]} = \frac{4K_D[E](K_2[X])^2}{1+4K_D[E](K_2[X])^2} \tag{50}$$

From Eq. (47) it is clear that:

$$\frac{d\ln[E]}{d\ln[X]} = \frac{2\dfrac{\bar{Y}_x}{1 - \bar{Y}_x}}{1+\dfrac{2\bar{Y}_x}{1 - \bar{Y}_x}} \tag{51}$$

Equation (42) therefore obtains the form:

$$n = 2 \, \frac{\overline{Y}_x}{1 - \overline{Y}_x} - \frac{\overline{Y}_x}{1 - \overline{Y}_x} \times \frac{2\dfrac{\overline{Y}_x}{1 - \overline{Y}_x}}{1 + \dfrac{2\overline{Y}_x}{1 - \overline{Y}_x}} \tag{52}$$

$$n = 2 - \frac{2\overline{Y}_x}{1 + \overline{Y}_x} \tag{53}$$

It can be immediately seen from Eq. (53) that n_H cannot exceed the value 4/3 at 50% ligand saturation. As pointed out in the text the maximal Hill coefficient of 2.0 is obtained at very low ligand occupancy (see text).

Case 2. Ligand binding only to monomer. Dimer predominant species in the absence of ligand.

The conditions require that $K_2 = 0$ and $8[E_o]K_D \gg 1$.

The expression for the Hill coefficient [Eq. (42)] takes the form:

$$\frac{d\ln \dfrac{\overline{Y}_x}{1 - \overline{Y}_x}}{d\ln[X]} = 1 - \frac{2K_D[E] \, \dfrac{d\ln[E]}{d\ln[X]}}{1 + 2K_D[E]} \tag{54}$$

introducing Eq. (51) into Eq. (54) one obtains:

$$n = \frac{d\ln \dfrac{\overline{Y}_x}{1 - \overline{Y}_x}}{d\ln[X]} = 1 + \frac{\overline{Y}_x}{2 - \overline{Y}_x} \quad \text{when } K_1[X] \gg 1 \tag{55}$$

Therefore the highest Hill coefficient at 50% saturation cannot exeed 4/3, exactly as in Case 1. In this case the value of the Hill coefficient increases as $\overline{Y}_x$ increases, as pointed out in the text.

2. Monomer Multimer Equilibrium Coupled to Ligand Binding

1. Association to a Multimer. Consider a protein with l subunits which associates to polymer of pl subunits:

$$pE_l \rightleftharpoons E_{lp} \qquad K_p = \frac{[E_l]^p}{[E_{lp}]} \tag{56}$$

and where binding occurs only to the polymer and therefore is infinitely cooperative:

$$E_{lp} + l_p X \rightleftharpoons (EX)_{lp} \qquad K = \frac{[EX]_p}{[E_{lp}][X]^{lp}} \tag{57}$$

The total concentration of monomer units, $[E_o]$, is given by:

$$[E_o] = l[E] + lpKK_p[E_l][X]^{pl} \tag{58}$$

The binding equation becomes:

$$\bar{Y}_x + \frac{lp \cdot KK_p[E]_l^{p-1}[X]^{pl}}{1 + lpK \cdot K_p[E]^{p-1}[X]^{pl}} \tag{59}$$

$$\frac{\bar{Y}_x}{1 - \bar{Y}_x} = p \cdot KK_p[E]_l^{p-1}[X]^{pl} \tag{60}$$

The Hill slope at any position is given by:

$$n = \frac{d\ln \dfrac{\bar{Y}_x}{1 - \bar{Y}_x}}{d\ln X} = pl + (p-1)\frac{d\ln[E_l]}{d\ln[X]} \tag{61}$$

From Eqs. (58) and (59) one can obtain:

$$\frac{d\ln[E_1]}{d\ln[X]} = -\frac{p l \bar{Y}}{1 + \bar{Y}_x (p-1)} \tag{62}$$

thus

$$\frac{d\ln \dfrac{\bar{Y}_x}{1 - \bar{Y}_x}}{d\ln[X]} = \frac{p l}{1 + \bar{Y}_x (p-1)} \tag{63}$$

evaluated at the midpoint gives:

$$n_H = \frac{2p l}{1 + p} \tag{64}$$

The ligand concentration at the midpoint is given by:

$$[X_{0.5}] = \left(\frac{1}{pKK_p}\right)^{\frac{1}{p l}} \left(\frac{2 l}{[E_o]}\right)^{\frac{p-1}{p l}} \tag{65}$$

or:

$$[X_{0.5}] = A \left(\frac{1}{[E_o]}\right)^{\frac{p-1}{p \cdot l}} \tag{66}$$

where

$$A = \left(\frac{1}{pKK_p}\right)^{\frac{1}{p l}} (2 l)^{\frac{p-1}{p l}} \tag{67}$$

2. Dissociation of a Multimer. Consider a polymer with m subunits, which is incapable of binding the ligand [X]. Binding only occurs when the polymer is dissociated.

$$E_m \rightleftharpoons mE \qquad\qquad K_m = \frac{[E]^m}{[E_m]} \tag{68}$$

100

$$E + X \rightleftharpoons EX \qquad K_1 = \frac{[EX]}{[E] \cdot [X]} \tag{69}$$

If very little free monomer is present at any time, the expression for the total concentration of monomers becomes:

$$[E_o] = K_1[E][X] + mK_m[E]^m \tag{70}$$

and

$$\bar{Y}_x = \frac{K_1[X]}{mK_m[E]^{m-1} + K_1[X]} \tag{71}$$

then

$$\frac{d\ln \frac{\bar{Y}_x}{1 - \bar{Y}_x}}{d\ln[X]} = \frac{d\ln \frac{K_1[X]}{mK_m[E]^{m-1}}}{d\ln[X]} = 1 - (m - 1)\frac{d\ln[E]}{d\ln[X]} \tag{72}$$

and

$$\frac{d\ln[E]}{d\ln[X]} = -\frac{K_1[X]}{m^2K_m[E]^{m-1} + K_1[X]} \tag{73}$$

or the Hill slope at any position is:

$$\frac{d\ln \frac{\bar{Y}_x}{1 - \bar{Y}_x}}{d\ln[X]} = 1 + \frac{(m - 1)\bar{Y}_x}{m(1 - \bar{Y}_x) + \bar{Y}_x} \tag{74}$$

When evaluated at the midpoint the Hill coefficient becomes:

$$n_H = 1 + \frac{m - 1}{m + 1} \tag{75}$$

Equation (75) is identical to Eq. (154) in the text.

References

Adair, G.S.: The hemoglobin system: VI. The oxygen dissociation curve of hemoglobin. J. Biol. Chem. 63, 529-545 (1925)

Adams, M.J., Blundell, T.L., Dodson, E.J., Dodson, G.G., Vijayan, M., Baker, E.N., Harding, M.M., Hodgkin, D.C., Rimmer, B., Sheat, S.: Structure of rhombohedral 2 zinc insulin crystals. Nature (London) 224, 491-495 (1969)

Barcroft, J., Hill, A.V.: The nature of oxyhemoglobin with a note on its molecular weight. J. Physiol. 39, 411-428 (1910)

Bohr, C., Hasselbach, K., Krogh, A.: Über einen in biologischer Beziehung wichtigen Einfluß, den die Kohlensäurespannung des Blutes auf dessen Sauerstoffbinding ausübt. Skand. Arch. Physiol. 16, 402-410 (1904)

Bränden, C.I., Eklund, H., Nordström, B., Bowie, T., Sodelund, G., Zeppensauer, E., Ohlsson, I., Adeson, A.: Structure of liver alcohol dehydrogenase at 2.9-Å resolution. Proc. Nat. Acad. Sci. 70, 2439-2442 (1973)

Caspar, D.L.D., Klug, A.: Physical principles in the construction of regular viruses. Cold Spring Harb. Symp. Quant. Biol. 27, 1-25 (1962)

Changeux, J.-P.: The feedback control mechanism of biosynthetic L-threanine deaminase by L-isoleucine. Cold Spring Harb. Symp. Quant. Biol. 26, 313-318 (1961)

Conway, A., Koshland, D.E.Jr.: Negative cooperativity in enzyme action. The binding of diphosphopyridine nucleotide to glyceraldehyde 3-phosphate dehydrogenase. Biochemistry 17, 4011-4023 (1968)

Cook, R.A., Koshland, D.E.Jr.: Specificity in the assembly of multisubunit proteins. Proc. Nat. Acad. Sci. 64, 247-254 (1969)

Cook, R.A., Koshland, D.E.Jr.: Positive and negative cooperativity in yeast glyceraldehyde 3-phosphate dehydrogenase. Biochemistry 9, 3337-3342 (1970)

Cornish-Bowden, A., Koshland, D.E.Jr.: The influence of binding domains on the nature of subunit interactions in oligomeric proteins. J. Biol. Chem. 245, 6241-6250 (1970)

Dahlquist, F.W.: Cooperativity in ligand binding to proteins showing association-dissociation phenomena: An application of the quantitative interpretation of the Hill coefficient. In: Methods in Enzymology. New York: Academic Press, in press

Dahlquist, F.W., Koshland, D.E.Jr.: Quantitative interpretations of the Hill plot (in preparation)

Derosier, D.J., Oliver, R.M., Reed, L.J.: Crystallization and preliminary structural analysis of dihydrolipoyl transsuccinylase, the core of the 2-oxoglutarate dehydrogenase complex. Proc. Nat. Acad. Sci. 68, 1135-1137 (1971)

Duncan, B.K., Diamond, G.R., Bessman, M.J.: Regulation of enzymatic activity through subunit interaction. A possible example. J. Biol. Chem. $\underline{247}$, 8136-8138 (1972)

Evarse, J., Kapaln, N.O.: Lactate dehydrogenase: Structure and function. Adv. Enzymol. $\underline{37}$, 61-133 (1973)

Fenselau, A.: Structure-function studies on glyceraldehyde-3-phosphate dehydrogenase. III. Dependency of proteolysis on NAD^+ concentration. Biochem. Biophys. Res. Commun. $\underline{40}$, 481-488 (1970)

Frieden, C.: Protein-protein interaction and enzymatic acitivity. Ann. Rev. Biochem. $\underline{40}$, 653-696 (1971)

Gerhart, J.C., Pardee, A.B.: The effect of feedback inhibitor, CTP, on subunit interactions in aspartate transcarbamitase. Cold Spring Harb. Symp. Quant. Biol. $\underline{28}$, 491-496 (1963)

Haber, J.E., Koshland, D.E.Jr.: The effect of 2,3-diphosphydyceric acid on the changes in β-β interactions in hemoglobin during oxygenation. J. Biol. Chem. $\underline{246}$, 7790-7793 (1971)

Hanson, A.W., Appelbury, M.L., Coleman, J.E., Wycoff, M.W.: X-ray studies on single crystals of *Escherichia coli* alkaline phosphatase (Appendix). J. Biol. Chem. $\underline{245}$, 4975-4976 (1970)

Heck, H. De A.: Statistical theory of cooperative binding to proteins. The Hill equation and the binding potential. J. Am. Chem. Soc. $\underline{93}$, 23-29 (1971)

Henis, Y.A., Levitzki, A.: The role of the nicotinamide and the adenine subsites in the negative cooperativity of coenzyme binding to glyceraldehyde-3-phosphate dehydrogenase. J. Mol. Biol. (1977)

Hill, A.V.: The combinations of hemoglobin with oxygen and with carbonmonoxide. Biochem. J. $\underline{7}$, 471-480 (1913)

Kirschner, K., Eigen, M., Bittman, R., Voigt, B.: The binding of nicotinamide-adenine dinucleotide to yeast D-glyceraldehyde-3-phosphate dehydrogenase: Temperature-jump relaxation studies on the mechanism of an allosteric enzyme. Proc. Nat. Acad. Sci. $\underline{56}$, 1661-1667 (1966)

Kirschner, M.W., Schachmann, N.K.: Conformational studies on the nitrated catalytic subunit of aspartate transcarbanylase. Biochemistry $\underline{12}$, 2987-2997 (1973)

Klotz, I.M., Langerman, N.R., Darnall, D.W.: Quaternary structure of proteins. Ann. Rev. Biochem. $\underline{39}$, 25-62 (1970)

Koshland, D.E.Jr., Némethy, G., Filmer D.: Comparison of experimental binding data and theoretical models in protein containing subunits. Biochemistry $\underline{5}$, 365-385 (1966)

Levitzki, A.: Ligand induced half-of-the-sites reactivity in rabbit muscle glyceraldehyde-3-phosphate dehydrogenase. Biochem. Biophys. Res. Commun. $\underline{54}$, 889-893 (1973)

Levitzki, A.: Half-of-the-sites reactivity in rabbit muscle glyceraldehyde-3-phosphate dehydrogenase. J. Mol. Biol. $\underline{90}$, 451-458 (1974)

Levitzki, A., Koshland, D.E.Jr.: Negative cooperativity in regulatory enzymes. Proc. Nat. Acad. Sci. $\underline{62}$, 1121-1128 (1969)

Levitzki, A., Koshland, D.E.Jr.: Role of an allosteric effector. Guanosine triphosphate activation in cytosine triphosphate synthetase. Biochemistry $\underline{11}$, 241-246 (1972a)

Levitzki, A., Koshland, D.E.Jr.: Ligand induced dimer-to-tetramer transformation in cytosine triphosphate synthetase. Biochemistry $\underline{11}$, 247-253 (1972b)

Levitzki, A., Koshland, D.E.Jr.: The role of negative cooperativity and half-of-the-sites reactivity in enzyme regulation.
In: Current Topics in Cellular Regulation. Horecker, B.L.,
Stadtman, E.E. (eds.). New York: Academic Press 1976, Vol. X,
pp. 1-40
Levitzki, A., Schlessinger, J.: Cooperativity in associating
proteins. Monomer-dimer equilibrium coupled to ligand binding. Biochemistry 13, 5214-5219 (1974)
Levitzki, A., Stallcup, W.B., Koshland, D.E.Jr.: Half-of-the-
sites reactivity and the conformational states of cytidine
triphosphate synthetase. Biochemistry 10, 3371-3378 (1971)
Matthews, B.W., Bernhard, S.A.: Structure and symmetry of oligomeric enzymes. Ann. Rev. Biophys. Bioeng. 2, 257-317 (1973)
Mockrin, S.C., Byers, L.D., Koshland, D.E.Jr.: Subunit interaction in yeast glyceraldehyde-3-phosphate dehydrogenase. Biochemistry 14, 5428-5437 (1975)
Moffet, F.J., Wang, T., Bridger, W.A.: Succinyl coenzyme A
synthetase of *Escherichia coli*. J. Biol. Chem. 247, 8139-8144
(1972)
Monod, J., Changeux, J.-P., Jacob, F.: Allosteric proteins and
cellular control systems. J. Mol. Biol. 6, 306-329 (1963)
Monod, J., Wyman, J., Changeux, J.-P.: On the nature of allosteric transitions: A Plausibe model. J. Model. Biol. 12,
88-118 (1965)
Moras, D., Olsen, K.W., Sabesan, M., Büchner, M., Ford, G.D.,
Rossman, M.G.: Studies of asymmetry in the three-dimensional
structure of lobster D-glyceraldehyde-3-phosphate dehydro-
genase. J. Biol. Chem. 250, 9137-9162 (1975)
Morino, Y., Snell, E.E.: The subunit structure of tryptophanase.
J. Biol. Chem. 242, 5591-5601 (1967)
Nichol, B.W., Jackson, W.J.H., Winzor, D.J.: A theoretical
study of the binding of small molecules to a polymerizing
protein system. A model for allosteric effects. Biochemistry
6, 2449-2456 (1967)
Novick, A., Szilard, L.: Dynamics of Growth Processes. Princeton,
N.J.: Princeton University Press 1954
Pauling, L.: The oxygen equilibrium of haemoglobin and its struc-
tural interpretation. Proc. Nat. Acad. Sci. 21, 186-191 (1935)
Perutz, M.F.: Stereochemistry of cooperative effects in haemo-
globin. Nature (London) 228, 726-734 (1970)
Perutz, M.F.: Nature of haem-haem interactions. Nature (London)
237, 495-499 (1972)
Rossmann, M.G., Adams, M.J., Buehner, M., Ford, G.C., Hackert,
M.L., Lentz, P.J., McPherson, A., Jr., Schevitz, R.W.,
Smiler, I.E.: Structural constraints of possible mechanisms
of lactate dehydrogenase as shown by high resolution studies
of the apoenzyme and a variety of enzyme complexes. Cold
Spring Harb. Symp. Quant. Biol. 36, 179-191
Rubin, M., Changeux, J.-P.: On the nature of allosteric tran-
sitions: Implications of non-exclusive ligand binding. J.
Mol. Biol. 21, 265-274 (1966)
Schlessinger, J., Levitzki, A.: Molecular basis of negative
cooperativity in rabbit muscle glyceraldehyde-3-phosphate
dehydrogenase. J. Mol. Biol. 82, 547-561 (1974)

Seydoux, F., Bernhard, S.A., Pfenniger, O., Payne, M., Malhotra, O.: Preparation and active-site specific properties of sturgeon muscle glyceraldehyde-3-phosphate dehydrogenase. Biochemistry 12, 4290-4300 (1973)

Steitz, T.A., Fletterick, R.J., Hwang, K.J.: Structure of yeast hexokinase. II. A 6 Å resolution electron density map showing molecular shape and heterologous interaction of subunits. J. Mol. Biol. 78, 551-561 (1973)

Teipel, J., Koshland, D.E.Jr.: The significance of intermediary plateau regions in enzyme saturation curves. Biochemistry 8, 4656-4663 (1969)

Umbarger, H.E.: Evidence for negative-feedback mechanism in the biosynthesis of isoleucine. Science 123, 848 (1956)

Valentine, R.C., Shapiro, B.M., Stadmann, E.R.: Regulation of glutamine synthetase. XII. Electron microscopy of the enzyme from *Escherichia coli*. Biochemistry 7, 2143-2152 (1968)

Valentine, R.C., Wrigley, N.G., Serutton, M.C., Irias, J.J., Utter, M.F.: Pyruvate carboxylase. VIII. The subunit structure as examined by electron microscopy. Biochemistry 5, 3111-3116 (1966)

Weber, G., Anderson, S.A.: Multiplicity of binding. Range of validity and practical test of Adair equation. Biochemistry 4, 1942-1947 (1965)

Wyman, J.: Linked functions and reciprocal effects in haemoglobin: A second look. Adv. Prot. Chem. 19, 223-286 (1964)

Wyman, J.: Regulation in macromolecules as illustrated by haemoglobin. Quart. Rev. Biophys. 1, 35-80 (1968)

Yates, R.A., Pardee, A.B.: Control of pyrimidine biosynthesis in *Escherichia coli* by a feed-back mechanism. J. Biol. Chem. 221, 575-770 (1956)

Subject Index

Molecular Biology, Biochemistry and Biophysics

Editors: A. Kleinzeller, G.F. Springer, H.G. Wittmann

Volume 1: J.H. van't Hoff
Imagination in Science
Translated into English with notes and a general introduction by G.F. Springer
1967. 1 portrait. VI, 18 pages
ISBN 3-540-03933-3

Volume 2: K. Freudenberg, A.C. Neish
Constitution and Biosynthesis of Lignin
1968. 10 figures. IX, 129 pages
ISBN 3-540-04247-1

Volume 3: T. Robinson
The Biochemistry of Alkaloids
1968. 37 figures. X, 149 pages
ISBN 3-540-04275-X

Volume 5: B. Jirgensons
Optical Activity of Proteins and Other Macromolecules
1973. 2nd revised and enlarged edition.
71 figures. IX, 199 pages
ISBN 3-540-06340-4

Volume 6: F. Egami, K. Nakamura
Microbial Ribonucleases
1969. 5 figures. IX, 90 pages
ISBN 3-540-04657-7

Volume 8
Protein Sequence Determination
A Sourcebook of Methods and Techniques
Edited by S.B. Needleman
2nd revised and enlarged edition. 1975.
80 figures. XVIII, 393 pages
ISBN 3-540-07256-X

Volume 9: R. Grubb
The Genetic Markers of Human Immunoglobins
1970. 8 figures. XII, 152 pages
ISBN 3-540-05211-9

Volume 10: R.J. Lukens
Chemistry of Fungicidal Action
1971. 8 figures. XIII, 136 pages
ISBN 3-540-05405-7

Volume 11: P. Reeves
The Bacteriocins
1972. 9 figures. XI, 142 pages
ISBN 3-540-05735-8

Volume 12: T. Ando, M. Yamasaki, K. Suzuki
Protamines
Isolation, Characterization, Structure and Function
1973. 24 figures, 17 tables. IX, 114 pages
ISBN 3-540-06221-1

Volume 13: P. Jollès, A. Paraf
Chemical and Biological Basis of Adjuvants
1973. 24 figures, 41 tables. VIII, 153 pages
ISBN 3-540-06308-0

Volume 14
Micromethods in Molecular Biology
Edited by V. Neuhoff
1973. 275 figures (2 in color), 23 tables.
XV, 428 pages
ISBN 3-540-06319-6

Volume 15: M. Weissbluth
Hemoglobin
Cooperativity and Electronic Properties
1974. 50 figures. VIII, 175 pages
ISBN 3-540-06582-2

Volume 16: S. Shulman
Tissue Specificity and Autoimmunity
1974. 32 figures. XI, 196 pages
ISBN 3-540-06563-6

Springer-Verlag
Berlin
Heidelberg
New York

Molecular Biology, Biochemistry and Biophysics

Editors: A. Kleinzeller, G.F. Springer,
H.G. Wittmann

Volume 17: Y.A. Vinnikov
Sensory Reception
Cytology, Molecular Mechanisms
and Evolution
Translated from the Russian by W.L. Gray
and B.M. Crook
1974. 124 figures (173 separate illustrations)
IX, 392 pages
ISBN 3-540-06674-8

Volume 18: H. Kersten, W. Kersten
Inhibitors of Nucleic Acid Synthesis
Biophysical and Biochemical Aspects
1974. 73 figures. IX, 184 pages
ISBN 3-540-06825-2

Volume 19: M.B. Mathews
Connective Tissue
Macromolecular Structure and Evolution
1975. 31 figures. XII, 318 pages
ISBN 3-540-07068-0

Volume 20: M.A. Lauffer
Entropy-Driven Processes in Biology
Polymerization of Tobacco Mosaic Virus
Protein and Similar Reactions
1975. 90 figures. X, 264 pages
ISBN 3-540-06933-X

Volume 21: R.C. Burns, R.W.F. Hardy
**Nitrogen Fixation in Bacteria
and Higher Plants**
1975. 27 figures. X, 189 pages
ISBN 3-540-07192-X

Volume 22: H.J. Fromm
Initial Rate Enzyme Kinetics
1975. 88 figures, 19 tables. X, 321 pages
ISBN 3-540-07375-2

Volume 23: M. Luckner, L. Nover, H. Böhm
**Secondary Metabolism
and Cell Differentiation**
1977. 52 figures, 7 tables. VI, 130 pages
ISBN 3-540-08081-3

Volume 24
Chemical Relaxation in Molecular Biology
Editors: I. Pecht, R. Rigler
1977. 141 figures, 50 tables. XVI, 418 pages
ISBN 3-540-08173-9

Volume 25
**Advanced Methods in Protein Sequence
Determination**
Editor: S.B. Needleman
1977. 97 figures, 25 tables. XII, 189 pages
ISBN 3-540-08368-5

Volume 26: A.S. Brill
Transition Metals in Biochemistry
1977. 49 figures, 18 tables. VIII, 186 pages
ISBN 3-540-08291-3

Volume 27
Effects of Ionizing Radiation on DNA
Physical, Chemical and Biological Aspects
Editors: A.J. Bertinchamps (Coordinating
Editor), J. Hüttermann, W. Köhnlein,
R. Téoule
1978. 74 figures, 41 tables. Approx. 420 pages
ISBN 3-540-08542-4

Springer-Verlag
Berlin Heidelberg New York

FSC
www.fsc.org
MIX
Papier aus verantwortungsvollen Quellen
Paper from responsible sources
FSC® C105338